LET THERE BE LIGHT

CENTER ON GLOBAL ENERGY POLICY SERIES

CENTER ON GLOBAL ENERGY POLICY SERIES
Jason Bordoff, series editor

Making smart energy policy choices requires approaching energy as a complex and multifaceted system in which decision makers must balance economic, security, and environmental priorities. Too often, the public debate is dominated by platitudes and polarization. Columbia University's Center on Global Energy Policy at SIPA seeks to enrich the quality of energy dialogue and policy by providing an independent and nonpartisan platform for timely analysis and recommendations to address today's most pressing energy challenges. The Center on Global Energy Policy Series extends that mission by offering readers accessible, policy-relevant books that have as their foundation the academic rigor of one of the world's great research universities.

Robert McNally, *Crude Volatility: The History and the Future of Boom-Bust Oil Prices*

Daniel Raimi, *The Fracking Debate: The Risks, Benefits, and Uncertainties of the Shale Revolution*

Richard Nephew, *The Art of Sanctions: A View from the Field*

Jim Krane, *Energy Kingdoms: Oil and Political Survival in the Persian Gulf*

Amy Myers Jaffe, *Energy's Digital Future: Harnessing Innovation for American Resilience and National Security*

Ibrahim AlMuhanna, *Oil Leaders: An Insider's Account of Four Decades of Saudi Arabia and OPEC's Global Energy Policy*

David R. Mares, *Resource Nationalism and Energy Policy: Venezuela in Context*

Agathe Demarais, *Backfire: How Sanctions Reshape the World Against U.S. Interests*

Johannes Urpelainen, *Energy and Environment in India: The Politics of a Chronic Crisis*

Let There Be Light

HOW ELECTRICITY MADE MODERN HONG KONG

Mark L. Clifford

Columbia University Press
New York

Columbia University Press
Publishers Since 1893
New York   Chichester, West Sussex
cup.columbia.edu
Copyright © 2023 Columbia University Press
All rights reserved

Library of Congress Cataloging-in-Publication Data
Names: Clifford, Mark, 1957– author.
Title: Let there be light : how electricity made modern Hong Kong / Mark L. Clifford.
Description: New York, NY : Columbia University Press, [2023] | Series: Center on global energy policy series | Includes bibliographical references and index.
Identifiers: LCCN 2022050841 (print) | LCCN 2022050842 (ebook) |
ISBN 9780231201681 (hardback) | ISBN 9780231201698 (trade paperback) |
ISBN 9780231554213 (ebook)
Subjects: LCSH: Economic development—China—Hong Kong—History—20th century. | Electrification—China—Hong Kong—History—20th century. | Electricity—Economic aspects—China—Hong Kong—20th century.
Classification: LCC HC427.9 .C6157 2023 (print) | LCC HC427.9 (ebook) |
DDC 338.95125—dc23/eng/20230201
LC record available at https://lccn.loc.gov/2022050841
LC ebook record available at https://lccn.loc.gov/2022050842

Printed in the United States of America

Cover design: Julia Kushnirsky
Cover image: Courtesy of the Hong Kong Heritage Project

Electricity is a marvelous thing:
We attach ourselves to the belief
That its instruments can remake
The whole world with a switch.

—WARWICK ANDERSON

CONTENTS

Chapter One
Private Light and Colonial Power   1

Chapter Two
In the Beginning: China Light & Power, 1900–1940   25

Chapter Three
War, Occupation, and New Possibilities: 1941–1946   67

Chapter Four
"A Problem of People": 1947–1958   89

Chapter Five
Electricity as a Political Project: 1959–1964   117

Chapter Six
"Die-Hard Reactionary" in the Expanding
Colonial State: 1964–1973   160

CONTENTS

Chapter Seven
"Intelligent Anticipation" for "1997 and All That": 1974–1982   192

Chapter Eight
Sing the City Electric   213

ACKNOWLEDGMENTS   225
NOTES   229
BIBLIOGRAPHY   261
INDEX   279

PHOTOGRAPHS FOLLOW PAGE 116

LET THERE BE LIGHT

Chapter One

PRIVATE LIGHT AND COLONIAL POWER

It was the stumble seen around the world. As British Prime Minister Margaret Thatcher left a lunch meeting at the Great Hall of the People in Beijing, she tripped and fell, ending up on her hands and knees at the bottom of the steps, as if she were kowtowing to the giant portrait of Mao Zedong across Tiananmen Square. That misstep on September 24, 1982, visually summed up a testing day for Thatcher: in the morning she had been told bluntly by paramount Chinese leader Deng Xiaoping that the resumption of Chinese sovereignty over the British Crown Colony of Hong Kong in 1997 was nonnegotiable.[1]

In Hong Kong itself, four days later, Thatcher inaugurated the first phase of what would become by 1989 one of the world's largest coal-fired electricity-generation plants, a facility whose construction resulted in the largest-ever export order for the British manufacturing industry. Ignoring the prospect of Hong Kong's now all too apparent return to China in 1997, Thatcher praised the Castle Peak facility as an "ideal opportunity for the British Government and British industry to work together and to produce a complete package for a project contributing to Hong Kong's continued expansion and development." She lauded the main backer, Lawrence Kadoorie, commending "his vision and his efforts."[2]

The trip to Beijing, the first to the Chinese capital by a serving British prime minister, and the ceremonial opening of the Castle Peak electricity plant, 1,200 miles to the south, would appear to have little in common. Thatcher's meetings in Beijing constituted international politics at the highest level, initiating negotiations for the surrender of the most significant remaining British colony. The Castle Peak visit celebrated a coal-burning industrial plant that provided electricity, by 1982 a basic everyday commodity in Hong Kong usually noticed only by its absence.

Yet the development of Hong Kong's electricity network was closely connected to the city's rapidly evolving political, economic, and social landscapes. China Light & Power (CLP) in turn was an actor that profoundly shaped the colony.[3] CLP's continuing expansion of its electricity-generating capacity in advance of demand (unlike cross-harbor rival Hongkong Electric, or HKE) ensured abundant supplies of electricity for the colony's fast-growing manufacturing sector and was a necessary prerequisite for Hong Kong's rapid economic acceleration in the post-1945 period. The evolution of Hong Kong's electricity system provides insight into the shifting relations between Hong Kong, Britain, and China in the four decades following World War II, a time that saw fundamental geopolitical and economic realignments across East Asia. The CLP plant inaugurated by Thatcher at Castle Peak in 1982 was designed from its inception with political considerations in mind. Lawrence Kadoorie (1899–1993), the company's charismatic chairman, had expressly placed orders with British suppliers as a way of keeping the British government engaged in the run-up to the expiration of Britain's lease of Hong Kong's New Territories on June 30, 1997.

Unlike Thatcher, Kadoorie had long assumed that the handover of the colony in 1997, if not earlier, was inevitable. The 1997 expiration of the New Territories lease did away with the need for China to take back the territory—forcibly or not—during the period of decolonization after 1945. The existence of the 1997 lease expiration extended Hong Kong's colonial era by several decades; this long colonial twilight provided a unique space for growth and experimentation in Hong Kong. It allowed the colony the luxury of prioritizing economic growth. Central to this ability to literally power industrial growth was the development of the electricity supply network.

Although Hong Kong's experience of electrification was unique, it shares some aspects with other places. Hong Kong's history of electrification may provide signposts for other places as they make energy choices, whether in their initial electrification or in the transition away from fossil fuels. These include, most notably, political struggles over who will pay how much and for what, that is, whether payments are for actual consumption (and if so, how much should reflect the fuel cost and how much the cost of fixed assets) or promised availability. The pricing of electricity is not straightforward, and some users invariably benefit from subsidies, whether they are implicit or explicit. Another issue is the inevitability of government involvement due to electricity's status at the time as a natural monopoly as well as its centrality to modern life and its technology- and capital-intensive attributes. This inevitability was seen starkly in the case of Hong Kong, where a self-described small state found itself pulled into key decisions about electricity, to the point where officials embarked on a lengthy and public exercise with the avowed aim of merging and nationalizing the colony's two electricity companies. A broader theme that Hong Kong shares with other places is the role that electricity plays in modernization; a related theme is the way in which the state uses the electricity supply system to extend its reach and control.

This look at Hong Kong challenges conventional colonial histories that tend to emphasize an elite politics and, when they mention business at all, stress Hong Kong's exemplary laissez-faire economic arrangements. Hong Kong, to quote a typical view, "succeeded by applying a generally obsolete, elsewhere discarded and thoroughly discredited economic system— 19th century *laissez-faire* capitalism."[4] In reality, as seen through the lens of the development of Hong Kong's electricity system, the colonial state and private enterprise were increasingly intertwined in the decades after 1945. This analysis suggests the need for a new approach to Hong Kong history that takes a more nuanced view of Hong Kong's much-commented-on laissez-faire economy. At the same time, any examination of the struggle to define the political contours of the electricity supply system highlights the need for students of Hong Kong to move beyond discussion of an elite political cadre to consider other actors and agents of change. Woven throughout this story is the role electricity played in economic development and the creation of a modern urban identity.

CLP was, along with Hongkong Electric, one of two electricity supply companies serving the colony. Each served geographically distinct areas: CLP, which was founded in 1900, provided electricity to Kowloon beginning in 1903 and, after 1928, the colony's New Territories; HKE, which was established in 1889 and started service in 1890, primarily served Hong Kong Island, the original site of the British Crown Colony as established by the

1842 Treaty of Nanking (Nanjing). This book focuses on CLP, the more dynamic of the two companies, and the one that consistently invested in new electricity-generating capacity in advance of its customers' needs.

The Hong Kong electricity supply configuration had a number of unusual aspects, namely the longstanding existence of two companies in a single city; the relatively small service areas of each; the absence of a formal regulatory structure; the lack of consolidation with larger regional players, due to Hong Kong's status as a colony; and the geographic distance from sources of fuel and manufacturers of generating equipment. These distinctive features, however, should not obscure the similarities with electricity supply companies in other urban centers around Asia and, indeed, around the world. The commonalities are more striking given what an outlier Hong Kong was in so many particulars. A study of the colony's electricity system provides critical insights that are helpful in thinking about the post-1945 development of electricity networks, industrial technologies, and economic development in other places both within Asia and throughout the non-European, non-Anglo-American world.

At the heart of this narrative is Lawrence Kadoorie himself, both as an actor—notably as the chairman of CLP—and as a commentator on Hong Kong's development. As head of CLP for five decades, Lawrence Kadoorie allows for a new examination of the complexities and contradictions of the British colonial laissez-faire state. Kadoorie was the son of a Baghdad-born Mizrahi Jew who had originally come to Hong Kong while working for the Sassoon family. The Sassoons pioneered the first Mizrahi businesses in India and China, opening the way for scores of successful Mizrahi families who achieved remarkable business success in the region, and in doing so they became among the world's wealthiest families. The Kadoories subsequently surpassed the Sassoons in wealth and influence. Although none of the Kadoories were notably religiously observant, their financial success in anti-Semitic Hong Kong opened them to attack and contributed to the 1959–64 attempt to curb CLP, of which the family owned almost 40 percent.

Lawrence Kadoorie was vociferously opposed to government actions in many areas. Moreover, one of his most primary interlocutors in the colonial government, Financial Secretary John Cowperthwaite, is known as the most important proponent of Hong Kong's laissez-faire capitalism. Yet the two men worked closely to jointly develop a system that subjected CLP to new and unprecedented government control. Far from stifling economic growth, these controls made it possible for CLP to secure Hong Kong's largest-ever foreign investment, from U.S. oil giant Esso (the forerunner of today's Exxon) in 1964, and later enabled the Castle Peak project that Thatcher opened. Those investments in turn provided the electricity that enabled Hong Kong to record one of the world's highest rates of sustained economic growth.

This account of state-company interactions is one that challenges conventional histories, where the emphasis tends to be on the success of Cowperthwaite's "positive non-interventionism" (where government intervention was held to be a last resort) and the victory of laissez-faire policies. In the case of the electricity supply companies, Kadoorie and Cowperthwaite were far less committed to any economic philosophy or ideological position than is generally assumed. Instead, they engaged in negotiated capitalism, with agreement on a quasi-regulatory system known as the Scheme of Control Agreement (SoCA), reached only after four years of intensive discussions.

Kadoorie is one of the most important individuals in the 156-year history of colonial Hong Kong. He played a pivotal role in building Hong Kong's electricity network and, as a result, the entire structure of the post-1945 city. Kadoorie and CLP made possible Hong Kong's economic acceleration and its emergence as a significant exporter. Because of decisions made by Kadoorie at CLP, the crown colony acquired importance in the 1950s and 1960s as a cheap-labor production site for manufactured exports and as a brightly lit outpost of capitalist modernity on the edge of monochromatic China.

Electricity exemplified modernity. Hong Kong's plentiful and, for practical purposes, virtually unlimited electricity stood in contrast to scarcity throughout Asia and, indeed, much of the world. The easy access to electricity gave those in the territory an identity as modern urban people and helped shape Hong Kong's identity as a new sort of colony. Electricity also enabled other manifestations of modernity, whether it be air-conditioned cinemas and well-lit shopping malls; elevators and escalators that enabled vertical movement in the territory's ever-taller buildings; tramways, trains, and a metro system; and the factories that allowed Hong Kong to become for a time one of the world's largest textile producers.

Kadoorie's outlook and his actions were, of course, also determined by the contexts of his time. The technologies he promoted were themselves the product of complex global networks. The large-scale systems that he built relied on many individuals as well as a constellation of interconnected technologies. They were far more complex than any one individual alone could fully understand, let alone manage. CLP's network has continued to grow, nearly three decades after Kadoorie's death. Today, CLP has significant investments in China and India and is one of the largest investor-owned power businesses in Asia.

Like most electricity supply companies, CLP consists of a large-scale system that in turn is part of a deep technological, business, political, and social power structure, one that is so interwoven with everyday life that it typically lies unexamined, hidden in plain sight. Electricity systems such as CLP's result from choices, both conscious and unconscious, both articulated and unarticulated. Despite the far-reaching political and social consequences of these choices, they are often written about in retrospect as if their development was inevitable and somehow natural. The development of CLP, and by extension Hong Kong, in the period between 1945 and 1982, was not in any sense preordained. Choices that defined the future of CLP and of the colony were made by Kadoorie and others at CLP.

Electricity systems are produced by—and in turn help to generate— interconnected political, social, and economic forces. Through the period from 1945 to 1982, key decisions were made in Hong Kong about how, where, and by whom the electricity system would be financed, manufactured, distributed, and used—decisions that had wide-reaching and long-lasting social, economic, and political effects. How electricity in Hong Kong was distributed and used was a product of the interplay between possibilities the technology allowed as well as a constellation of social, political, economic, and business decisions that were both unique to Hong Kong and at the same time echoed similar choices and possibilities in other colonies and urban milieus.

ELECTRICITY AS SYSTEM

Humans create technology. Technology, in turn, shapes the possibilities open to individuals and to societies. Technology itself is embedded in a web of administrative, legal, social, business, and other arrangements. Large-scale electricity supply systems such as CLP's lend themselves to centralized, hierarchical systems operated by skilled managers.[5] In Hong Kong there was a dynamic three-way interplay between individuals (first among them Lawrence Kadoorie), institutions (especially the colonial government and CLP, but also social institutions including newspapers and civic reform associations), and the technological system of electricity.

Electricity systems are also complex assemblages that reflect different agendas, both private and public. Technologies like electricity are embedded in a collectively held vision of how a society ought to be ordered, with attention to the dynamics of power and control that are inherent in these relationships.[6]

Electricity was one of the most important new technologies of the twentieth century. The widespread application of electrical technology in the form of both light and machine power transformed societies and economies during this time. It was also the core component of extensive and complex technological systems.

The manufacture, transmission, and distribution of electricity constitutes one of those rare inventions known as a "macroinvention" or a "general purpose technology."[7] These foundational technologies have, in the words of technology scholar David Nye, profound and widespread ramifications across "all aspects of the economic system." Technologies like the internal combustion engine and the Internet "differ from niche technologies because they have so many uses. They also have considerable scope for improvement, many distinct applications, and strong complementarities with other technologies." Significantly, they can be combined with other technologies to produce new possibilities and outcomes.[8] Electricity is worthy of study because it makes so much else possible.

To be useful for humans, electricity must be manufactured. In contrast with water, naturally produced electricity is found only in erratic and limited forms, most commonly lightning strikes or the discharge of static electricity. Unlike water, which falls from the sky and can be collected and used, useable electricity is always manufactured. Electricity can only be stored in limited circumstances, so it must generally be distributed and consumed as soon as it is produced.

The physical phenomenon of electricity is of less interest here than the systems that were constructed to produce, distribute, and use it and the influence they exerted on political, economic, and social developments in Hong Kong and elsewhere. Electricity systems comprise networks of power in a literal, physical sense. Global satellite images of the Earth at night show a concentration of light in economically prosperous areas, with a strong correlation between the intensity of light and an area's economic wealth. Electricity produces wealth, and wealth produces a demand for yet more electricity.

CLP's electricity system includes the construction of the supply network, notably the generating station, the high-voltage transmission network (both under city streets and on large metal towers in rural areas), and the distribution wires that by 1982 had been extended to transmit current into most rooms in the colony's urban areas.

HONG KONG AND TECHNOLOGY

Hong Kong was never a center of innovation. Although CLP was not responsible for technological invention, in the sense that it is usually understood (the development of a new device or process), it was a technology-rich company due to its significant technological recombination. Recombination involves the use of existing equipment and techniques in novel ways. The difficulty of operating often unstable electricity supply technology in a colony halfway around the world from equipment manufacturers (and operating in difficult weather conditions) posed continuing technical challenges that required adaptation, innovation, and applied expertise on the part of CLP technical staff, managers, and executives. The challenges of providing reliable electricity supplies in CLP's service area, comprising both some of the world's densest urban settings and remote mountain villages in a subtropical environment prone to typhoons, were significant.

Electric light was introduced to the colony in 1890, just eight years after Thomas Edison set up the first distributed electricity supply systems in London and New York. Hong Kong was an early adopter but not a pioneer. The colony was ahead of Singapore and Dublin in introducing electric light but behind Shanghai and Bombay. Electricity-generating and transmission equipment and technology were brought to Hong Kong because of personal and institutional links to Britain and Germany by the founders of HKE, and subsequently by CLP founder Robert Shewan and later by Kadoorie. Hong Kong companies and their wealthier consumers were willing and able to pay for up-to-date technology. CLP itself had as one of its core principles—one that was formally articulated by Lawrence Kadoorie—a commitment to buying the best and most up-to-date equipment.[9] This embrace of the most modern technology is evident in the CLP archives; in the case of electricity generation, this includes oil-, coal-, and nuclear-powered electricity-generating technology.

A willingness to invest in building additional electricity-generating capacity while looking ahead to new technologies that would make these same investments obsolete is striking; this approach reflects Kadoorie's vision, but it also exemplifies a more general ethos prevailing in Hong Kong that embraced modernity as expressed through technology. The relative openness to new technologies that had been developed elsewhere allowed Hong Kong a high degree of success in recombining technologies. To take one example, the combination of electricity, used to make ice, and the internal combustion engine, which replaced sails on traditional fishing boats, had a powerful effect on the local fishing industry, on Hong Kong diets, and the internationalization of the Hong Kong economy.

Hong Kong's lack of natural resources hampered CLP. The ability to secure supplies of coal, the fuel used for the electricity-generating system in Hong Kong during much of the period under consideration, was quite different in the colony than it was in resource-rich countries such as Great Britain, Germany, France, and Japan. These countries all benefited from plentiful supplies of easily accessible coal to power their electricity plants. In Hong Kong, all coal was imported. Coal's price, timely availability, and quality were a source of recurrent anxiety.[10] For a time, there was similar anxiety over securing oil supplies.

All generating and transmission equipment, technology-rich products such as VHF radios and computers, and even most of CLP's senior managers, also needed to be imported. That Hong Kong had a sophisticated electricity system despite having to import virtually all the parts of this system, reflects the considerable ability of Kadoorie and other members of the colony's elite to recombine an array of systems.

For goods that needed to be physically transported, Hong Kong was a long sea voyage away from the scientific, educational, financial, and political center in Britain. The opening of the Suez Canal in 1869 was the most important transportation innovation affecting Hong Kong until the coming of regular air transportation after the end of World War II. After the extension of the telegraph to Hong Kong in 1871, a congratulatory telegram sent from London to the Hong Kong General Chamber of Commerce took fifty-three minutes to be received.[11] Hong Kong was simultaneously far and near.[12] Information could travel in hours or days but, for much of colonial Hong Kong's history, goods took weeks or months to arrive. After the record-breaking foreign investment from Esso in 1964, which saw the oil giant take a significant ownership stake in CLP's new Tsing Yi electricity plant, Hong Kong was increasingly anchored to the United States.

One should be careful not to overemphasize Hong Kong's relative technological sophistication and CLP's success in combining the different elements that made up its electricity supply system. Late deliveries and missing or damaged parts were a chronic complaint until the mid-1960s when the Esso investment and the growing affordability of air travel and air shipments collapsed distance and ameliorated problems of installing and operating equipment at sites so far from equipment manufacturers.

Personal and institutional relationships with technology- and capital-rich multinational petroleum companies aided CLP's acquisition of technology and associated material resources. Shell supplied the fuel oil that powered CLP's boilers before Esso's investment in 1964, and Kadoorie originally hoped to bring the company into a joint venture with Esso.[13] Mobil, which had a petroleum refinery in Lai Chi Kok, in CLP's service area, negotiated with Kadoorie from November 1962 until July 1963 about the possibility of using some of its land for a venture with CLP to build a new electricity-generating station.[14] At the same time, beginning in July 1962, Kadoorie started talks with Esso.[15] Esso ultimately invested in this generating plant in order to secure CLP as a large and long-term customer for its oil. Kadoorie in turn personally nurtured the relationship with senior Esso executives, spending significant amounts of time with them in New York and elsewhere. He used his network to research a variety of technologies relating not only to electricity generation but also to water treatment and desalination.

Kadoorie's parlaying of a relationship with a single Esso representative into a transformative investment by one of the world's largest oil companies was part of a pattern that saw the Kadoorie family tap into the highest reaches of economic and political life in Britain, China, and the United States. Kadoorie's and CLP executives' relationships with Chinese officials allowed the company to realize a long-held dream of selling electricity to China beginning in 1979, and subsequently of constructing the country's first civilian nuclear reactor.

In the case of the Kadoories and CLP (as well as Hongkong Electric), these electric power networks were mostly developed with Britain. This is unsurprising, given that Hong Kong was a British Crown Colony, but worth noting since Kadoorie later made use of his political and business network in Britain. Most of CLP's electricity supply equipment came from Britain. Its expatriate managers were almost exclusively British, many of them having worked elsewhere in the British Empire.

When the colony wanted to establish a regulatory framework for electricity supply companies, two of the three members of the 1959 Electricity Supply Companies Commission (ESCC) came from Britain. The recommendation they proposed to merge and nationalize the colony's two electricity supply companies reflected the political arrangements prevailing at the time in Britain. Hong Kong ultimately decided to accept British technology in the form of electricity-generating stations, transmission lines, and other physical components but rejected merger and nationalization. In the late 1970s, Lawrence Kadoorie used connections with Britain's civil service and political leaders to win support from Prime Ministers James Callaghan and Margaret Thatcher for what became Britain's largest-ever manufacturing export order, an order that was paid for in part with concessionary financing provided by the British government.

Lawrence Kadoorie and CLP in many senses inverted the traditional relationship between the Crown Colony and Britain. It was Kadoorie, coming from an increasingly wealthy Hong Kong, who from the 1950s through the 1980s provided the industrial orders that protected the manufacturing jobs British ministers so desperately wanted to keep. Kadoorie repeatedly reminded the British public of the profit opportunities Hong Kong offered. In 1981, he was named the first Hong Kong–born peer of the realm, Baron Kadoorie of Hong Kong and of the City of Westminster, because of the Castle Peak electricity project's support for British manufacturing.[16]

The development of this technology- and capital-intensive system in Hong Kong has added significance because of the colony's unique geopolitical position. Hong Kong acted as an airlock between the vastly different systems of the communist People's Republic of China (PRC) and the capitalist world led by the United States and Britain. Hong Kong's colonial period extended four to five decades longer than those in other significant colonial territories because the June 30, 1997, expiration of the New

Territories lease provided a convenient endpoint for both the People's Republic and Britain, obviating the need to negotiate an earlier resumption of sovereignty. The extended colonial period in turn allowed for the rapid economic growth. Powerful geopolitical and geo-economic forces brought to bear on post-1945 Hong Kong produced and made visible elements of power, in all its senses, that often remain hidden, allowing us to see social, political, economic, and cultural forces that worked through and were in turn influenced by CLP's electricity system.

Hong Kong challenges conventional narratives of electricity supply companies, especially those in colonial settings and other areas outside of European and Anglo-American economies, in several ways. First, Hong Kong's electricity supply companies were never nationalized; second, they have long-lived corporate structures (these are some of the world's oldest electricity supply companies under continuous management); and third, both have emerged in recent decades as sizable global investors.[17] Hong Kong was also unusual in having not one but two electricity supply companies. Both CLP and HKE coupled a high degree of local management autonomy with an ability to use global finance and technical and managerial expertise.[18]

ELECTRICITY AND THE LIMITS OF HONG KONG LAISSEZ-FAIRE

CLP was a privately owned company in a city that prided itself on having one of the least interventionist governments in the world. Many commentators have claimed that Hong Kong during British colonial days represented the post-1945 world's best example of a laissez-faire economy. Nobel Prize–winning economist Milton Friedman extolled the city as the "modern exemplar of free markets and limited government." He singled out Financial Secretary John Cowperthwaite for his adherence to laissez-faire policies. Just weeks before Friedman's own death in 2006, Friedman wrote that Cowperthwaite

> was so famously laissez-faire that he refused to collect economic statistics for fear this would only give government officials an excuse for more meddling. His successor, Sir Philip Haddon-Cave, coined the term "positive non-interventionism" to describe Cowperthwaite's approach. The results of his

policy were remarkable. At the end of World War II, Hong Kong was a dirt-poor island with a per-capita income about one-quarter that of Britain's. By 1997, when sovereignty was transferred to China, its per-capita income was roughly equal to that of the departing colonial power, even though Britain had experienced sizable growth over the same period. That was a striking demonstration of the productivity of freedom, of what people can do when they are left free to pursue their own interests. . . . Whatever happens to Hong Kong in the future, the experience of this past 50 years will continue to instruct and encourage friends of economic freedom. And it provides a lasting model of good economic policy for others who wish to bring similar prosperity to their people.[19]

In Cowperthwaite's dealings with CLP, the financial secretary was not in fact as "famously laissez-faire" as Friedman would have us believe. Cowperthwaite was Lawrence Kadoorie's principal interlocutor in the Hong Kong government's attempts to nationalize, merge, and regulate CLP. His actions in the electricity debate of the early 1960s were more in line with the confiscatory policies that Friedman so abhorred than the laissez-faire worldview with which Friedman credited the Hong Kong financial secretary.

Cowperthwaite displayed a cavalier and even dismissive regard for CLP shareholders and their property rights. He was financial secretary during most of the nearly five-year period when the government imposed a ceiling on CLP's dividend payments, and he personally and repeatedly threatened nationalization to pressure Kadoorie into accepting controls over electricity tariffs and new investments. During this period, Cowperthwaite attempted to force a merger of the colony's two electricity companies on the grounds of fairness to consumers and business efficiency and insisted that the government be allowed to appoint two directors, including the first chairman. He maintained that a substantial portion of the revenues paid by electricity consumers belonged to those consumers and not to CLP, and that they should be used to reduce rates.

The attempted nationalization of CLP and the related effort to merge it with HKE following the release of the ESCC's recommendations in 1960 were at odds with the laissez-faire insistence that the state should take only the most limited role in business and economic organization.

This abortive state takeover was significant historically because it reflected the context of its time and involved two of the most important and capital-intensive enterprises in the colony; furthermore, its main protagonists were two of the best-known and powerful individuals of their time.

Despite the oft-repeated claim that the colony excelled in limited government, CLP worked closely with government authorities, including both civilian administrators and the British military, to electrify the city. The intimate relationship between the colonial administration and CLP was sometimes so close that the two organizations, viewed purely on a functional basis, could sometimes be distinguished only with difficulty. This was especially true after 1960, when CLP adopted government pay grades and pay scales for its employees' wages and salaries. The change mirrored CLP's emerging status as a sort of parallel state, taking on some of the responsibilities that governments shouldered in many jurisdictions. Beginning in the 1950s, CLP's annual reports refer to the role it played in supporting numerous government projects, such as installing and maintaining street lights, providing electricity to squatter resettlements and semiofficial low-cost housing schemes, as well as the redevelopment of old sites and construction of private houses.[20]

Government officials and CLP employees worked together in many ways. At the system level, after 1964 they jointly negotiated decisions on large-scale capital investments and profitability arrangements. At the furthest end of the distribution system, where electricity is delivered to a plug in a consumer's home, CLP worked with the government to provide electricity for squatter settlements as a way of ensuring public safety and social order. At another terminal point in the distribution system, CLP collaborated with the government to implement a program of installing and operating the street lighting network in Kowloon and the New Territories. At a third end point, CLP supported the government's network of water reservoirs by building electricity supply used in pumping water, including an emergency public works department pumping station between Shing Mun reservoir and Tai Po market. Government administrators often mediated between CLP and villagers in negotiating for the provision of electric lighting to more remote areas.[21] The hybrid corporate-state creation that Cowperthwaite and Kadoorie were instrumental in constructing in the "famously laissez-faire" colony shows the impossibility of hands-off government policies in a world of complex technologies like electricity supply. This intertwining of corporate and state interests demonstrates the economic, political, and social importance of energy and electricity, even in small-government Hong Kong.

A careful examination of the political, social, technological, and economic choices made at this time in Hong Kong reveals the interweaving of public and private spheres, and it suggests that we would do well to question long-held assumptions about Hong Kong's economic environment. Far from a triumph of laissez-faire capitalism, the expansion of Hong Kong's electricity supply system was the product of a state-private joint effort whose high-growth years were from 1945 to 1982, the period under review. This was a period that later commentators said represented the high-water mark of Hong Kong's successful laissez-faire economic policies.

Hong Kong was not in any sense unusual in seeing state involvement in electricity supply arrangements. Large-scale electricity systems everywhere are a product of political and social choices as much as economic ones. Thomas Hughes, a historian of electricity supply systems, highlights the powerful interests that inevitably build and maintain large-scale systems: "A century ago Karl Marx pointed out the ways which vested interests, especially capital, shape the course of history. In fact, large technological systems represent powerful vested interests of another kind."[22] The politicization of electricity supply arrangements, where these arrangements were widely debated and contested, typically occurred during the initial decades of electrification in Europe and the United States.[23] In Hong Kong, it was not until almost seven decades after the introduction of electricity in the colony, from 1958 until 1964, that electricity was embraced as an explicitly political subject.

ELECTRICITY AND THE IDEA OF HONG KONG

CLP was integral to the making of post-1945 Hong Kong, in ways that have not previously been understood. Viewing the city through the prism of its fast-expanding electricity network helps show the physical, political, and economic structure that was created in this period. Moreover, the symbiotic relationship between CLP and the colonial state undercuts the notion of Hong Kong as a laissez-faire economy—a place of paradigmatic free trade and minimal economic regulation. In fact, CLP used the imposition of government controls over its operations as a source of strength, transforming these restrictions into an implicit state guarantee of profitability that allowed the company to grow in scale and sophistication. CLP's rapid increase in electricity generation capacity played a pivotal role in the colony's ability to achieve a degree of self-determination and international standing; the manufactured exports, which required reliable and abundant supplies of electricity, allowed Hong Kong to accumulate significant sterling reserves that in turn gave the colony a high degree of autonomy.

More generally, CLP helped to redefine how Hong Kong people experienced and thought of their city, and how the world looked at Hong Kong and its people. The colony emerged for British manufacturers as both a threat, because of its inexpensive manufactured goods (especially textiles) that were manufactured with CLP's electricity, and as an opportunity, notably as a sizeable market for capital goods, such as those used in the Castle Peak power plant and other electricity-generating plants, and in CLP's transmission and distribution network. The global scale of Hong Kong's resources and international-standard capabilities added new texture, and new tensions, to the traditional relationship between Britain and its increasingly wealthy colony.

Urban life in the period from 1945 to 1982 in Hong Kong was characterized by the increasingly widespread availability of affordably priced electrical power. The dramatically increased volume of electrical current sold by CLP during the almost four decades under consideration occurred at a time when several important transitions were underway. The colony's population increased almost ninefold during the 1945–1982 period, intensifying the demand for the government to provide more and better housing, education, health, water, and public order.

Electricity created a new kind of city, from the neon signs that came to characterize the colony to the fans and ice and air-conditioned cinemas. People used electric current to power their lights, refrigerators, and rice cookers.[24] These new devices changed people's experiences of what it meant to be modern, to be a Hong Kong person, and confirmed its residents' belief that they were living in the most modern place in China. CLP played a significant role not only in powering factories and lighting shops but also in shaping the very idea of the city. CLP offered hope of a better future, a world powered by electricity. The demand for electric current produced revenues that enabled CLP to build yet more electricity-generating supply, which in turn made possible increased electricity use.

In 1946, CLP had 26,000 consumers. A few months before Thatcher opened the Castle Peak plant in 1982, Lawrence Kadoorie presented a videotape recorder to the company's one millionth customer. That represented almost a fortyfold increase in customers in less than four decades. Hong Kong's electricity-fueled modernization paralleled what was happening in many other countries and territories globally, although it happened faster and more dramatically in Hong Kong than in most places. The more than hundredfold increase in CLP's generating capacity, from 19.5 megawatts (MW) in August 1945 to 3,006 MW at the end of 1982, was extremely high by any standards.[25] Electricity sales grew by five hundred times in less than four decades, from eighteen million kilowatt hours (kWh) in 1946 to almost nine billion kWh. Per capita electricity use rose more than a hundredfold, from 20 kWh to 2,201 kWh.[26]

During this period, the public's attitude toward electricity shifted dramatically. Electricity was no longer viewed as a luxury good. It acquired a new status as a basic commodity that everyone should have access to at a reasonable price. That shift in public expectations forced concessions and accommodations by CLP and HKE.

Electricity during most of the twentieth century was considered a natural monopoly, a business with high startup costs that rewards early entrants who are able to stay financially solvent until they are profitable. Hong Kong, almost uniquely among significant global cities, did not treat the electricity supply as a monopoly. There were—and are—two distinct electricity supply companies, although they served geographically distinct areas of the colony. No franchise agreement existed between the government and the electricity companies, notwithstanding a loose nonexclusive framework for part of CLP's service area. By global standards, the absence of state involvement in electricity supply was noteworthy and reflects Hong Kong's long tradition of minimal government involvement in business. How and why that changed after 1945, particularly in the 1959–1964 period, is the focus of this book.

CLP, before World War II a distant second in size to HKE, began to increase its capacity as soon as the war was over. Much of this increase was to power the thousands of factories that were established in the colony by entrepreneurs from Shanghai and elsewhere. One estimate says that as many of two-thirds of Shanghai's cotton spinning factories moved to Hong Kong.[27] Lawrence Kadoorie was one of the people most responsible for the emergence of Hong Kong as a manufacturing center. Shanghai textile manufacturers moved to CLP's service area because Kadoorie personally promised them as much electricity as they needed.

After 1945, Kadoorie lobbied Shanghai textile entrepreneur Y. C. Wang (Wang Yuncheng) to relocate his operations to Hong Kong with promises of abundant electricity at a time when Shanghai businessmen were considering moving to places like Brazil and Mauritius.[28] Kadoorie subsequently sold Wang land for his factory and served as chairman of Wang's Nanyang Textiles. Wang later joined CLP's board. The textile industry was critical to Hong Kong's newfound export prowess. It employed tens of thousands of people, and it used large amounts of electricity seven days a week. Although there was significant manufacturing in the colony before 1941, these new factories allowed Hong Kong to shed its entrepôt status and become an export-oriented manufacturing-based economy.

The migration of many of the Shanghai elite to Hong Kong brought some of that city's flavor to what had been, in the words of Governor Alexander Grantham, "a small village" in contrast to "the great cosmopolitan center" of pre-Communist Shanghai.[29] The appeal of Hong Kong was all the stronger given political persecution and economic mismanagement in China after the establishment of the PRC. The failure of China's Great Leap Forward (1958–1962) led to one of the deadliest famines in human history, killing an estimated thirty to forty-five million people.[30] As China turned in on itself, the colony's businessmen and workers—all of them consumers, too—transformed Hong Kong from a second-rank colonial port to a metropolis that almost by default inherited much of Shanghai's chic image.

Hong Kong officials worried about "a problem of people," as about 1.5 million people, mostly refugees from China, settled in the colony between 1945 and 1960. It was not electricity but manual labor that predominated in mid-1950s Hong Kong, as a young immigrant from Shanghai remembered: "A small window, closed against the full afternoon glare, looks out on to a granite slope where, hour after hour, day in and day out, men, women and even children may be seen breaking stone with mallet and chisel. Above them, squatter huts scab a swathe of hillside. So thick is the swarm of people on that incline that it is like gazing at a kicked ant hill."[31]

The image of Hong Kong started to change. While this was a period of water shortages and, in 1967, bombs and the threat of incipient revolution, Hong Kong was also a city of energy, of electric energy and human energy—one feeding off the other. A government-sponsored promotional film about the city released in 1961 contrasted the "gentle landscape of the New Territories" to Hong Kong's gritty urban milieu. Footage of water buffalo working in rice paddies was juxtaposed against the "dazzling variety of new industries" and burgeoning residential areas where people were "more densely concentrated than in central London or Manhattan." The Hong Kong people, concluded the narrator, "ask of the world only the freedom to work and with it the dignity and self-respect of paying their way."[32]

Electricity transformed what colonial administrators termed the "problem of people" into an opportunity for economic growth. This boom was jump-started in a way the world had never seen: inexpensive labor combined with modern technology—including plentiful supplies of electricity—and financial and human capital to produce what would later be called the East Asian miracle, and Hong Kong was a leading example.[33] It was a new world, and electricity production was one of the preconditions for its birth.

Electricity enhanced human energy and allowed for increased economic growth. Improved economic output in turn raised incomes, thus leading to more consumption, which in turn led to demands for more electricity in a feedback loop that would continue for more than four decades. CLP's electricity transformed the nature of labor in the colony, making widespread factory work possible. That opened up new economic possibilities, including the ability to bring women into the industrial workforce in unprecedented numbers.

Vladimir Lenin famously declared that Soviet power plus electrification equaled Communism.[34] Kadoorie and Hong Kong provided a more successful alternative: electrification plus capitalism equaled high economic growth, social mobility, and fewer political tensions. Electricity allowed the colony to bridge the gap between its disparate parts, between rural and urban, between traditional and modern.

Hong Kong's material and economic separation from China after 1949 forced it to become more geographically diversified and more international in its economic activities. The new People's Republic of China cut many economic ties. Then two Korean War embargoes were imposed on China by the United States and by the United Nations in 1950 and 1951, respectively. Though it remained dependent on China for vital goods such as food and water, its focus shifted to the West, with the United States taking on unprecedented importance.

The appearance of Esso as CLP's joint venture partner in 1964 was part of Hong Kong's growing dependence on the United States. Capital, technology, music and movies, and news and ideas from the United States, Britain, and elsewhere in Asia flowed into and circulated freely within the colony. Ideas that emerged closer to home about the future of China spread too, although the colonial government restricted their free dissemination so that Hong Kong would not become a battleground between Communists and Nationalists. British authorities further restricted civil and political liberties after the violence of the PRC's Cultural Revolution spread to the colony in 1967.

Hong Kong emerged in the collective global imagination as one of Asia's—indeed the world's—most alluring cities. Hong Kong mixed the soupçon of rare and exotic goods from off-limits China—items such as sea slugs, caterpillar fungus, silks, and ceramics—with the order and predictability of British colonial rule. All these elements were thrown together in a city built against the spectacular natural backdrop of steep hillsides rising from its famous harbor.

This history of electricity connects geopolitics to a world of small businesses and "ordinary" people. For Margaret Thatcher and Deng Xiaoping and Lawrence Kadoorie, wrestling as they did with issues of national prestige and sovereignty and economic development, the electricity-generating stations and their distribution networks literally embodied centralized power. At a more personal level, electricity had a profound impact upon daily life: on fishermen and restaurant owners who relied on electrically powered refrigeration; on factory workers whose lives hummed to a new machine-driven rhythm; on students who now studied at night; and on families watching television accompanied by the cooling whirr of electric fans. Electric lights at home and in offices and retail shops created new opportunities but also imposed new controls on work, school, and even sleep patterns.

In so doing, electricity fueled a nervous, restless energy whose pace was amplified by, and in turn further amplified, urban Hong Kong life. By 1982, the majority of households and almost every business in Hong Kong were connected. Cables and wires linked the most intimate commercial and domestic spaces, penetrating into bedrooms and bathrooms, boardrooms and private offices. In this sense, electricity was a unifying technology whose impact was felt throughout society and whose power, literally and figuratively, opened up new, more modern, possibilities. Such was the positive, dynamic feedback loop of this foundational technology.

Even before the advent of electricity, light was associated with order and the state; darkness connoted crime, fear, and disorder. Historian Wolfgang Schivelbusch has written about the social importance of light and reactions against light, with revolutionary lantern breakers hanging hated members of the ancien regime from symbolically important light poles in the revolutionary France of the 1790s. In Hong Kong, mandatory night passes had from the earliest days of the colony in the 1840s regulated the movement of Chinese residents and helped keep the anxiety and fear generated by darkness at bay for colonists who lived in an overwhelmingly Chinese-populated city.

Histories of nineteenth-century Hong Kong make much of its lawless reputation; nighttime added to the fear of robbery and burglary.[35] A curfew imposed in October 1842 prohibited Chinese, except watchmen, from being out after 11:00 p.m.[36] Soon Chinese were required to carry a lantern after dark; lanterns in Hong Kong served to light the way but, as Schivelbusch notes of lantern use in Western Europe, their main purpose was to make the bearer visible and thus to lessen the chances of a surprise attack. "Anyone who did not carry a light," he writes, "was regarded as suspect and could immediately be arrested, like someone without papers."[37] Although the same requirement was initially imposed in Hong Kong, by 1847 Governor John Davis reported that "the Police of the Town was improved and it was no longer necessary to enforce the regulation for the Chinese inhabitants to carry lights at night."[38] Still, after 9:00 p.m., all Chinese needed to carry a letter from their employers authorizing them to be out of doors.[39]

The pass system was problematic, with an 1857 government order complaining that night passes had been "abused" and instituting measures to "restrict very considerably the number of Night Passes hitherto issued."[40] The requirements for lamps, lanterns, and night passes make it no surprise that the first use of gas and electrical lighting in Hong Kong, as in many other cities, was to light public streets and prevent criminals from roaming the city under cover of darkness.

Schivelbusch's work on Paris demonstrates that early lamps generated little illumination, and Hong Kong's pre-electric lights share this characteristic. These lamps did little more than symbolize the desire for order and safety; they were meant to reassure the colony's law-abiding residents that the lawlessness and violence lurking in the subtropical colonial night could be kept at bay.[41]

The idea of electricity as a medium of control was ubiquitous. Cobras, as early traffic signals were known, directed traffic. More accurately, they instructed motorists to control themselves and their motorized vehicles. What traffic policeman had done to control drivers was now done by an impersonal electricity-dependent agent.[42] Elevators and escalators, too, disciplined and regulated physical movement in and across the city. The order imposed by factory machines on the rhythms of human life is a well-worn theme; an electrified regime controlled hundreds of thousands of Hong Kong workers in the 1950s and 1960s.

CLP and HKE extended physically into more Hong Kong households than any other organization except the government. The 1898 treaty granting Britain a ninety-nine-year lease on the New Territories exempted a small military outpost, later known as the Kowloon Walled City, from British control. After the Communists came to power in China, the enclave was nominally subject to the PRC's sovereignty and thus off-limits to colonial officials. CLP nonetheless provided electricity, and its meter readers were admitted into the area.

Crime-fighting and the overall imposition of order implied by street lighting remained a preoccupation for both CLP and the government through the 1950s. In the 1970s, the extension of the electricity network to squatter areas aimed both to improve living conditions and to foster order.[43] CLP's role in bringing not only electricity but also order and control was apparent in these informal settlements, where the threat of fire and disease preoccupied colonial authorities.[44] A well-regulated electricity system aided public health by promoting a more modern and regulated lifestyle. Electricity was clean and modern, in contrast to smoky and easily combustible kerosene.[45]

There was a broader geopolitical context to Hong Kong's electricity project in the post-1949 period, when its very survival as a British colony was in doubt. In 1957, reflecting recurrent worries about the possibility of a takeover by the PRC, U.S. President Dwight Eisenhower secretly agreed to defend Hong Kong in the event of a Chinese attack.[46] Hong Kong's colonial boundaries did not altogether protect the territory from the Chinese civil war, with Nationalist (Kuomintang, or KMT) and Communist (Chinese Communist Party, or CCP) forces occasionally resorting to violence in the colony. The most notable incident occurred in October 1956, when some sixty people died in Hong Kong's deadliest riots after the removal of a KMT flag during celebrations of Taiwan's national day.[47]

Commentators dubbed Hong Kong the "Berlin of the East." This label was overly dramatic and misleading, since Hong Kong had access to Chinese supplies of food and water, whereas Berlin had been isolated by the Soviet blockade. But Hong Kong's large refugee population, coupled with ongoing political uncertainty within the colony and in the PRC, made this a city on the edge—not a place to invest in assets that could not be moved, such as electricity supply systems.[48]

Hong Kong was a place of light set against the drab and dark backdrop of Maoist China. To justify Eisenhower's promise to defend Hong Kong, the city had to be developed as a visible alternative to communism. A Colonial Office "profit and loss account" prepared at the request of British Prime Minister Harold Macmillan noted that Hong Kong "provides a demonstration adjacent to a Communist country of what can be done by good administration combined with rule of law and respect for individual freedom."[49] The bright lights of everything from jewelry shop windows to bars and restaurants signified capitalist abundance, in stark contrast to the enforced poverty of a benighted China. CLP and HKE both implicitly recognized the appeal of the life they made possible, showcasing the colony's well-lit streets and electrified signs in annual reports and other company materials.[50]

Light served more blatantly political purposes in times of crisis. As a gesture of defiance to Chinese Communists and a symbol of faith in Hong Kong during the 1967 Cultural Revolution riots and bombings, Lawrence Kadoorie ordered construction crews to work around the clock at his new St. George's headquarters building in Central, which was under construction at the time, "so that the sparks from welding machines could illuminate the night sky and inspire confidence in the future of Hong Kong."[51] CLP's service area in the New Territories and Kowloon was a forward base in this campaign of capitalist modernization. It was in CLP's territory that most refugees first set foot as a result of the colonial government's "touch-base" policy, which gave them the right to remain in Hong Kong if they registered in an urban area of the colony.

CLP provided the current for the new towns such as Sha Tin and Yuen Long that were built to deal with fire-prone informal settlements. The new towns were yet another example of the dynamic interplay between electricity, the growth of the city, the imposition of social order, and the introduction of new possibilities. CLP played a role in the successful implementation of Hong Kong's large-scale public housing program, begun after fire destroyed the Shek Kip Mei squatter area in 1953, by designing an electricity supply system specifically for the new housing estates.[52] CLP sold the electricity to the film studios that lit their movie sets and thus created and transmitted a vision of capitalist, consumerist success and abundance. (The anticommunist message was aided by funding for Hong Kong's Asia Pictures by the CIA-backed Committee for a Free Asia.[53]) By 1961, the faster-growing CLP sold almost twice as much electricity as HKE, providing the current that showed capitalist Hong Kong could provide for people in a way that Communist China could not.[54]

CLP's reliance on electrical demand from factories distinguished Hong Kong from U.S. cities, notably Chicago, but also from leading European cities including London, Paris, and Berlin, where growth in consumer demand was more important. Electricity charges almost quadrupled from prewar levels when military administration ended in mid-1946, as part of an agreement between the electricity companies and military authorities. An opaque fuel surcharge introduced in 1950, designed to pass on fuel costs to consumers, prompted public unhappiness. For the first time, electricity became the subject of intense political debate.

Hong Kong was not unique. During times of rapid growth, "tension between the utilities and political institutions such as local government was high," Thomas Hughes notes in his account of the development of electricity systems in Europe and the United States.[55] In Europe and the United States, this tension was most apparent before World War II, and in some cases before World War I. In Hong Kong the tension climaxed between 1959 and 1964. After unprecedented public hearings, the Electricity Supply Companies Commission in early 1960 recommended merging and nationalizing the colony's two electricity companies. To win compliance, the government adopted an unprecedented policy: it restricted dividends, controlled tariffs, and exercised oversight over CLP's capital spending plans. Ultimately, CLP successfully resisted both merger and nationalization. An examination of the company's history shows what made that resistance possible.

Chapter Two

IN THE BEGINNING

China Light & Power, 1900–1940

Dwarfed by a massive Union Jack and two flanking Republic of China flags, Hong Kong Governor Geoffrey Northcote and CLP Chairman Lawrence Kadoorie inaugurated the company's largest electricity-generating facility inside a large boiler hall at CLP's Hok Un (Hok Yuen) facility on the Kowloon peninsula, on February 26, 1940, before a crowd of some 1,500 invited guests.

The Hok Un expansion, which increased generating capacity by two-thirds, represented the most emphatic mark that the Kadoories had made on CLP since taking control of the company almost a decade earlier. Northcote's presence at the Hok Un opening signaled the importance of this new, self-consciously modern electricity plant and the boost that it would give Kowloon and Hong Kong. His appearance and Kadoorie's speech also marked the first significant public recognition of the Kadoories' leading role as controlling shareholders of CLP.

Electricity station openings and expansions often inspire public displays of civic pride, and the Hok Un celebration constituted part of this tradition.[1] This celebration appears to have been the first of its kind in Hong Kong. The opening ceremony was the first time that a Hong Kong governor had officiated at a CLP plant event and the first recorded event of any kind to mark one of the company's plant expansions. Similarly, HKE has no record of previous electricity-related inauguration ceremonies.

Northcote and Kadoorie consciously sought to frame the event as a historic one. In their speeches, they looked back to the past of a barren and undeveloped Kowloon, one whose soil was supposedly too poor even to grow cabbages, and contrasted it with a present that was full of promise. This depiction of precolonial and early colonial Hong Kong as undeveloped echoed the thinking of British Foreign Secretary Lord Palmerston in 1841, who rebuked Superintendent of Trade Charles Elliot for accepting Hong Kong Island as settlement during the first Opium War. Hong Kong was "a barren island with hardly a house upon it," complained Palmerston in relieving Elliot of his position. "You have treated my instructions as if they were waste paper."[2] A century later, Kadoorie and Northcote extolled the new Hok Un electricity generation station's establishment as a tangible investment in the future prosperity of Kowloon, the colony's less developed urban area that bordered the northern edge of Hong Kong harbor.

Kowloon was, from a colonial view, comparatively little developed at the beginning of the twentieth century. It was home to British military troops, the Hong Kong and Whampoa Dock Co., and a small number of other industrial businesses, as well as warehouses and wharves. Kowloon had a raffish reputation. Lawrence Kadoorie remembered that new arrivals before World War II were asked: "Are you married or do you live in Kowloon?"[3] The quip hinted at Kowloon's reputation for attracting those who wanted to conceal debt, scandal, or extramarital, and often interracial, liaisons.

Kowloon's development was given a boost with the 1898 acquisition of the New Territories by the British from the Qing government on a ninety-nine-year lease. This added the area north of Boundary Street to the colony, turning Kowloon into an important node. Kowloon lay at the tip of a peninsula of mainland China and, with the addition of the New Territories, provided significant opportunities for development. The construction of the Kowloon-Canton Railway, which opened in 1911, connected the colony with mainland China and further bolstered the confidence of those who believed in Kowloon.

By the time of the electricity station's opening in 1940, Kadoorie and Northcote's faith in historical progress had been tempered by war.[4] During the six years of planning and construction for the Hok Un expansion, the world had moved toward what soon would become total war. Japan's invasion of China, which had started in 1931 with the seizure of Manchuria and continued for the next six years at a low level, escalated into a large-scale conflict in 1937.

The Japanese Imperial Army invaded Guangdong province in October 1938. On February 21, 1939, nine Japanese planes crossed into Hong Kong airspace as part of a bombing raid on the mainland border town of Shum Chun (Shenzhen). Colonial authorities instituted mandatory registration of British subjects in April. The first compulsory blackout drill took place on July 27. On August 16, the Kowloon-Canton railway was closed at the border "owing to the presence of Japanese troops." Press censorship began on August 26. On September 1, Germany invaded Poland. Britain declared war on Germany on September 3. German nationals in Hong Kong were interned in LaSalle College, and the liquidation of German firms began on September 8.[5]

The Hok Un generating station was built with war in mind. It included an air raid shelter and could be blacked out quickly. The implicit message in Northcote's and Kadoorie's speeches was that the colony had prevailed in the face of adversity in the past and would do so again.

The Asian war and the European wars merged with the Japanese attack on Pearl Harbor on December 7, 1941, and the entrance of the United States into the conflict. Japanese troops attacked Hong Kong on December 8, just a few hours after the assault on Pearl Harbor.

China had already become a major theater of war thanks to Japan's 1937 invasion. The impact on Hong Kong was significant. By 1940, the colony had been transformed by an influx of more than 750,000 Chinese refugees fleeing the fighting that accompanied the Japanese invasion. The refugees had almost doubled the colony's population in the three years following the Japanese invasion of China. (The 1931 census recorded 840,473 residents.) An estimated 100,000 Chinese refugees came to the colony in 1937; in 1938 some 500,000 moved across the border. Another 15,000 came in 1939.

When war began in Hong Kong, the colony's population was estimated unofficially by air raid wardens at 1.64 million.[6] An estimated half-million people were sleeping on the streets.[7] Efforts were made to set up refugee camps and to offer inducements for repatriation.[8] Government attempts to limit the influx of refugees were ineffective, prompting antigovernment sentiment that infected the wartime internment camp at Stanley.[9]

In the 1930s, CLP was transformed from a peripheral company producing a patchily distributed product to one that, with the assertion of management control by the Kadoorie family and the decision to invest in the Hok Un expansion, put in place the physical, managerial, and, it should be stressed, ideological underpinnings that allowed the electricity supply company to become an indispensable actor in the post-1945 industrial takeoff and urban expansion.[10] CLP was transformed from an agency business run out of the Shewan Tomes & Co.'s office in the St. George's Building in Central to a more capital-intensive and professionally managed business. The transformation was manifested concretely by the opening of CLP's Bauhaus-inspired head office on Argyle Street in Kowloon in 1940.

Most significant treaty ports and urban colonial areas in the Asia-Pacific region had a distributed electricity supply network by 1900, less than twenty years after the concept of distributed electricity was pioneered in Britain and the United States. These included Bombay, Calcutta, Canton, Hong Kong, Manila, Melbourne, and Shanghai, as well as Beijing and Tokyo. Relative latecomers in this context, such as Singapore and treaty ports in China, generally built their distributed electricity supply networks no later than the first decade of the twentieth century.[11]

The relatively rapid introduction of electricity generation and distribution technology demonstrates both the recognized usefulness of this technology and the global networks of knowledge transmission that existed in the late nineteenth century. The rapid adoption of new technologies in so many colonies, among other places, suggests that their adoption was, if not inevitable, highly probable in places with strong connections to London and other metropolitan centers, where there were sufficient financial and human resources to take advantage of technological innovations.

The development of capital- and technology-intensive electricity supply systems in the colony was made possible because of the broader international context of globalization in the late nineteenth century, including the rapid diffusion of up-to-date technologies and techniques to geographically dispersed points: "While long-distance trade by land and by sea is an ancient phenomenon," note historians Jürgen Osterhammel and Niels Petersson, "integrated world markets for goods, capital, and labor were unknown before the middle of the nineteenth century."[12]

Technology, too, became globalized. Edison's 1882 demonstrations of distributed electricity supply systems in London and New York were replicated in dozens of cities around the world within a decade. These systems spawned novel inventions that became everyday technologies, including elevators, which allowed for taller buildings and denser cities; refrigeration, which both permitted longer food supply chains and changed domestic shopping and cooking patterns; and lighting, which changed domestic, work, educational, and consumption patterns.

Electrification, and the establishment of colonial electricity grids, must be seen in part as elements of a system of mapping, surveillance, control, and the imposition of power. Beyond this obvious sense of the grid as an instrument of power, electricity also brought opportunities and liberation. Electricity brought lights to city streets and the dwellings of the affluent, extending each day's time for working and socializing and increasing the feeling of safety; electricity pumped water, making cities healthier. The long-distance movement of both fresh and waste water meant that larger cities could be built. These innovations required larger-scale networks of finance, technological adeptness, and managerial organization to be effectively exploited.[13]

The process of globalization accelerated in Asia between 1860 and 1880 with the extension of telegraph service from Europe and the opening of the Suez Canal in 1869. These two events occurred around the same time as the introduction of important technological improvements in steamship engine and propeller efficiency. Steamships using the Suez Canal made travel from Europe to India and China faster and more reliable; travel along the China coast and up the Yangtze River into the country's heartland, and to China's treaty ports, became regular and routine. The transcontinental telegraph, which came to Hong Kong in 1871, eroded the tyranny of distance and made communication between significant Asian cities and international centers such as London, Paris, and New York a matter of days rather than months.[14]

By the end of the nineteenth century, the widespread diffusion of electrification and of the internal combustion engine constituted important new components of technological globalization. The human networks that brought electricity supply technology to Asia were primarily organized by a small number of Europeans, Americans, and British, as well as entrepreneurs who operated under British protection, such as the head of the Kadoorie family in Asia, Eleazer ("Elly") Silas Kadoorie, and his son Lawrence.

TECHNOLOGY IN THE COLONY

Hong Kong was always technologically modern in the sense that the city's elite, both Chinese and otherwise, fairly quickly and consciously adopted many commercially available products and technologies developed in Europe and the United States. Hong Kong was adept at combining and assembling relatively complex systems using components imported from abroad, notably from Britain. These components included both the engineering and technical elements we traditionally associate with technology, as well as capital, or what we now call financial engineering.

With lighting, Hong Kong followed a trajectory that was similar to that seen in Melbourne, Shanghai, Bombay, and Singapore.[15] Gas lighting began in the colony in December 1864 when the London-based Hong Kong and China Gas Co. began street lighting.[16] HKE turned on the colony's first electric street lights in 1890. CLP was founded in 1900. Its stated aim was to produce electricity in Kowloon, but its first act was to take over an electricity-generating plant in Canton; it started generating electricity in Kowloon only in February 1903.[17]

Although the colony's elite inhabitants adopted technologies, often rapidly, there are no records of research laboratories or of individuals who were regarded as inventors. The University of Hong Kong, the colony's first university, was established in 1911. Medicine and engineering were two of its three core subjects (arts was the third), yet the university had little in the way of research libraries, let alone research capacity; there were concerns in the British business community in Hong Kong as well as in the Advisory Committee on Education in the Colonies in London about the lack of qualified engineers, with the university subjected to withering criticism.[18]

CLP was typical in buying technology and expertise from elsewhere. Its archives have no record of internal research capabilities during this period; the company relied on its own staff's applied technological expertise combined with more fundamental knowledge and guidance from equipment suppliers and consultants. The physical distance from equipment manufacturers was a source of ongoing complaints by CLP management, at least through the 1950s. The company seemingly took no steps to overcome its lack of research capacity, although in 1957 it funded a vocational training program that ultimately became part of Hong Kong Polytechnic University.[19]

CLP ensured that it benefited from its foreign suppliers' technical expertise. A 1937 letter from British engineering consultants Preece, Cardew & Rider outlined measures to guarantee the integrity of an electricity meter during shipment from England:

> Arrangements are being made to have the equipment concerned repacked [after testing at the University of Manchester] specially for export at Manchester and dispatched direct to Hong Kong from the College. The kilowatt-hour meter after testing will be sealed and placed in the care of the Ship's Purser during shipment. No doubt arrangements can be made locally to receive this instrument from the ship at Hong Kong to prevent the possibility of damage due to careless handling at the port.[20]

This reliance on Britain (and, more generally, Europe and the United States) for technology, and as a technological reference point, enabled the rapid adoption of new technologies. Governor George William Des Voeux commented on the brilliant lighting and noted that "the gas-devices were quite as good as the best in London."[21] Local telephone service started in 1882, just six years after Alexander Graham Bell's patent registration.[22] Julius Reuter set up an office of his eponymous news agency in Hong Kong no later than 1871, though Reuters appears to have had a presence in the colony at least as early as 1862, the same year that Hong Kong was connected by telegraph, via Singapore and Bombay, with London.[23]

The cross-harbor Star Ferry inaugurated its services in 1888, the same year that the Peak Tram, Asia's first funicular, began operations.[24] The first movie theater opened in 1896, and department stores made their appearance in 1900. Robert Shewan, the founder of both CLP and the large Shewan Tomes trading and agency firm, which for many years ran the electricity supply company, inaugurated operations of electric streetcars in 1904. The automobile first came to Hong Kong in about 1905—CLP Consulting Committee member Joseph Noble was said to have the colony's first private automobile—and buses arrived in 1909. The Kowloon-Canton Railway, whose twenty-two-mile New Territories and Kowloon section was reputed to be "one of the most expensive pieces of railway construction in the world," began operations in 1910, making it theoretically possible to travel by train from Hong Kong to Calais.[25]

The first airplane arrived in 1911, although it would be 1936 before the Kai Tak airport began commercial passenger operations, with an Imperial Airways flight linking Hong Kong with Britain via Penang.[26] In October, Pan American World Airways began flights across the Pacific via Manila, and China National Airways Corp. provided an air link to the mainland.[27] Radio broadcasts in the colony began in 1928.[28]

Colonial Hong Kong had long excelled in engineering and construction technology, skills that would later underpin its successful electrification. Its susceptibility to heavy rains and typhoons prompted the colony to invest in higher-quality infrastructure, including roads. Reflecting on his time in Hong Kong, Des Voeux, who served as governor from 1887 to 1891, noted, "The roads were admirably constructed and kept, those ascending the Peak being entirely of Portland cement, so as to withstand the torrential rains, which would quickly have swept away any less durable material."

A storm in May 1889 dumped thirty-six inches of rain on the colony in thirty-two hours, but, other than washing away part of the Peak Tram tracks, caused only minor damage, testifying to "the exceptional solidity of the materials used in the construction of both public and private works."[29] The governor professed himself to be "struck with the enterprise which had created such a city on such a spot," given the need to cut building foundations and roads out of the solid granite bedrock that underlay the city.[30] This infrastructure development was almost exclusively on Hong Kong Island.

Hong Kong developed expertise in creating land from the ocean. With the thriving city of Victoria (today's Central) hemmed in by steep hillsides to the south, the harbor to the north, and military-controlled Admiralty to the east, reclamation was an expedient option. A large-scale scheme was approved in 1889. The Praya Reclamation project was led by Paul Chater, an ethnic Armenian who had been born and raised in Calcutta but had made his fortune in Hong Kong. Although various individual reclamation projects had been undertaken since the colony's founding, Chater's fifty-seven-acre plan was the largest yet. It comprised a seventy-foot-wide esplanade, or praya, that stretched almost two miles (3,051 yards) along a new foreshore. Financing was a hybrid of public and private, with owners paying the costs of reclamation proportional to the shorefront of their lots.[31]

In 1890 Prince Arthur, the Duke of Connaught, laid the foundation stone for the Praya Reclamation. The project, which was completed in 1904, was significant for its importance in the physical shaping of Hong Kong. Today the Court of Final Appeal, Mandarin Hotel, Prince's Building, St. George's Building, Alexandra House, Chater House, and the Hong Kong Club are all on land that was created as part of this project. Statue Square, Connaught Road, Des Voeux Road, and the Cenotaph are also on the Praya Reclamation. The reclamation was important both as a source of wealth and as an undertaking that knit together key members of the colony's non-Chinese elite, among them Parsis, Armenians, Jews, and Scots.

Hong Kong was also an important source of funding to build Chinese railways, again demonstrating the coupling of money and technological infrastructure development. In 1898 the Hongkong and Shanghai Bank (today's HSBC) and Jardine, Matheson formed the British and Chinese Corporation to provide the capital required for any rail concessions obtained from China. The corporation was given the right to build the Kowloon-Canton Railway in that year.[32] The completion of the railway in 1910 stimulated economic activity and development in Kowloon, CLP's service area.

Railroads during the late nineteenth century, like electricity-generating networks of the twentieth century, required significant amounts of financial resources, technological expertise, and skilled labor. Italian workers who had been working on the Yunnan railway in Indochina were brought in to construct the Kowloon-Canton railway, and Indian surveyors were imported to map the New Territories because they were deemed more expert than their Chinese counterparts.[33]

A different illustration of Hong Kong's part in an increasingly technologically sophisticated empire was the new HSBC headquarters. The building, the bank's third headquarters at the One Queen's Road site, opened in 1935, in the midst of the Great Depression. The bank's instructions to architects Palmer and Turner were to "please build us the best Bank in the world." There was more at work than provincial pride. The new bank structure was the tallest building between San Francisco and Cairo, and among the most technologically advanced, with elevators, air conditioning, and central heating—all of these technologies significant consumers of electricity. It boasted the colony's fastest elevators, twelve automatic ones that traveled at five hundred feet per minute, or almost six miles per hour.[34]

The architects needed to convince Chief Manager Vandeleur Grayburn that air-conditioning, which had been introduced to the colony only in 1931, was necessary; in the end Grayburn was apparently influenced by a report from Shanghai Commissioner of Public Health J. H. Jordan on the air-conditioning system in use at the bank's Shanghai branch: "My own impression is that air-conditioning on the whole tends to diminish disease and certainly renders the staff very much more comfortable and more able to work efficiently throughout their periods of duty. I am also of the opinion that the fact that air-conditioning is growing all over the world indicates the need for some such procedure."[35]

There are several noteworthy points about Jordan's remarks. First, there is the claim that air-conditioning promotes health; second, that it increases efficiency; and third, that Hong Kong and Shanghai must be mindful of and keep pace with global trends in new technologies. Although most members of the Hong Kong elite, non-Chinese and Chinese alike, prided themselves on the colony's technology-enabled connections to the outside world, they displayed no enthusiasm for the political or social reforms—notably higher wages and labor unions—that were underway in Britain or elsewhere. Popular political movements were repeatedly tamped down.[36] Technology could be imported from Britain and elsewhere, but ideas of political reform or social progress would be kept away as much as possible.

ELECTRICITY IN COLONIAL ASIA: HONG KONG, CANTON, AND BEYOND

Electricity generation had been progressively mastered in the decades following Michael Faraday's 1821 discovery that current could be produced by rotating a wire around a stationary magnet. The telegraph, which throughout the nineteenth century was generally powered by electric batteries, represented the most significant commercial application of electric technology.

The widespread fascination with electricity was, as noted, given a powerful practical and business impetus with Thomas Edison's demonstrations in London and New York in 1882 that electricity could be produced at a central power station and distributed to nearby homes and businesses for the purposes of lighting. Distributed electricity, along with the lightbulbs that Edison and Joseph Swan had independently invented and then sold jointly, allowed the rapid development of electricity supply companies.[37] The ability to economically generate and then transmit and distribute electricity over a large geographic area remained, for decades, both a significant technical challenge and a lucrative business opportunity for electricity supply companies.

Electricity-generating technology became more widely diffused throughout the 1880s, reflecting both the importance of this macroinvention and the speed with which technologies could be spread globally given the nineteenth-century improvements in transportation and communication. Electricity first appeared in Hong Kong aboard visiting ships. In his account of the 1887 celebrations held to mark Queen Victoria's Golden Jubilee, Des Voeux noted the presence of "some fifty ocean steamers, twenty to thirty other vessels and innumerable junks and other vessels, outlined with electric or other lights." He observed that the overall effect "made a spectacle which could hardly be equaled outside of fairyland," but his offhand remark about electricity suggests that the technology on ships anchoring in Hong Kong was by 1887 a fairly common occurrence.[38] This passing comment contrasts with the more extensive treatment Des Voeux gave to electricity in the Tokyo hotel room in which he stayed four years later. Contrasting incandescent light bulbs with the harsh arc lights often used for illumination, he wrote, "All the rooms were lit with the electric light, the bedrooms having in them that luxurious form of it which is only now to be seen in the best European hotels[.] I mean that which enables comfortable reading in bed, brightening one's book and at the same time protecting the eyes from glare."[39]

Melbourne, Calcutta, and Bombay were among the earliest sites in colonial Asia and Australia where electricity was produced. Electric light was first generated in Melbourne in the late 1860s and a nighttime football match was played in August 1879. The Victorian Electric Company was formed in 1880 and supplied arc lighting to the Eastern Market. In Calcutta, the first demonstration of electric light was conducted in 1879. In 1881, thirty-six electric lights were put into operation at the city's Mackinnon & Mackenzie Co. cotton mill. Bombay's Crawford Market was lit by electricity in 1882, although the lighting supply company went into liquidation two years later and the market was again illuminated by gas until the early 1890s.[40]

Electricity as a technology was revolutionary. As a business, electricity supply companies often struggled. Profitability was often elusive due to unstable technology, a narrow base of end-users, and the large and continuing capital investments required to achieve the economies of scale needed to reduce prices and thus get more consumers. For a time in the post-1945 period, investments in the stocks of electricity supply companies were so safe and their dividends so reliable that they were deemed fit for, in Wall Street parlance, "widows and orphans." In these early decades they were more akin to dotcom or cryptocurrency investments of the twenty-first century—shares subject to wild speculation and, often, corporate failures that saw shareholders suffer total losses.

The Shanghai Electric Co. was established in 1882. Like many early electricity ventures, it struggled to survive and was reorganized in 1888 as the New Shanghai Electric Co. This venture, too, ran into difficulties, and in 1893 it was taken over by the Municipal Council of the International Settlement in Shanghai.[41] A significant part of the Council's public works budget was devoted to the electricity network. This public ownership of the electricity supply and generating system was unusual, if not unique, in Asia in the 1890s. Shanghai also had an electric tramway and extensive gas lighting operations.[42] By the first decade of the twentieth century Dairen (Dalian), Hankow (Hankou, now part of Wuhan), Tientsin (Tianjin), and Tsingtao (Qingdao) either had electricity or, in the case of Hankow, were building electricity supply plants.[43]

Hong Kong was one of many cities in China where foreigners invested in new technologies. Foreign investment interests dominated China's electricity generation and supply sector, accounting for 1.5 megawatts (MW) of the 2.7 MW of the country's total generating capacity in 1911.[44] Steam-powered electricity generation in turn-of-the century China was only one of the ways that steam power was used. Its use extended beyond the boundaries of the colonial enclaves and treaty ports: "Flour mills, cotton spinning mills, steel works, collieries, waterworks, and electric light works are springing up, not only in the vicinity of the Treaty ports but throughout the country, for the Chinaman of today is almost as familiar with steam-power as is his European contemporary."[45] This statement suggests that to understand when and how technology such as steam power was disseminated from colonial enclaves into China more broadly, and how it was controlled and operated by Chinese, will require further research into China's early adoption of industrial technology.

The Calcutta Electric Supply Corp., founded in London in 1897, began service in 1899. The city's tramways abandoned horses and switched to electricity in 1902.⁴⁶ Electricity came to the Philippines in 1890; Manila's electricity company, La Electricista, was founded in 1892; a successor company, the Manila Electric Railroad and Light Company (later Meralco), was founded in 1903 to provide both electricity and tramway services.⁴⁷

Other colonial cities in Asia set up electricity supply companies in this period, often in conjunction with electric tramways. Tramways helped balance the demand for electricity throughout the day, an important consideration in the early decades of electricity supply when expensive generating equipment was idle for most of the day and in demand mostly during the early evening hours when lighting was required. Large customers such as tramways, shipyards, and cement works that contracted to buy significant amounts of current allowed the generating units to be run nearer to their capacity for more hours of the day, spreading out capital costs over a larger volume of electricity and thus reducing the cost per kilowatt hour (kWh) that electricity supply companies needed to charge. Thus, bulk industrial and transportation customers were typically granted significant discounts to the per-unit prices paid by lighting customers, especially residential ones. These rate differentials were to be a significant point of contention between CLP and its critics in the late 1950s and early 1960s.

Hong Kong businessmen were quick to see the potential for providing light. Hongkong and China Gas's Hong Kong head, Phineas Ryrie, applied to the colonial government as early as 1882 for a concession to supply electric light.⁴⁸ His request was countermanded by the company's board in London.⁴⁹

Hong Kong's first electric supply company, HKE, was incorporated in January 1889. Paul Chater was among its six directors. The company, which sought to sell 300,000 shares at $10 each, promised that "light for light" its electricity would be cheaper than gas. (Throughout, all currency is HK$ unless noted.) Directors claimed: "The advantages of ELECTRIC LIGHT are so well recognized that but little need be said on the subject." They nonetheless wrote that electricity was especially well suited to Hong Kong because electric light, unlike gas lamps, was not affected by heat or wind. HKE noted that it had purchased a site for its electricity-generating station and was ready to order the "most advanced" machines from England. The company added that its electricity would primarily be used for light but that it was also negotiating with the government for a contract to pump water from the Albany pumping station.⁵⁰

Half of the share issue was privately subscribed, presumably for the most part by Chater, other directors, and their associates. The other half was sold to the public, making it one of Hong Kong's first publicly traded stocks. The company noted in its advertisement that it hoped to have as many of the shares as possible in public hands. This may have reflected unhappiness that the gas company was organized in London and few shares were sold in Hong Kong. Author and former colonial official Austin Coates contends that, were it not for the backlash against London interests following the gas offering, it would have been simpler to organize the project in London, closer to the machinery manufacturers.[51]

Hongkong Telegraph editor and owner Robert Fraser-Smith crowed that the founding of companies such as HKE augured well for Hong Kong and heralded a new era of "science and civilization," one symbolized by "the blaze of electricity" that would metaphorically and physically bring light to China:

> The increase of these local public undertakings [such as HKE] is a sure pledge of our future prosperity both as a British trading emporium and as the pioneer of social progress in South China. Science and civilization irradiate from this rocky island over the neighboring populous areas of Chinese life and activity, and with the increase in the facilities of inter-communication, we may safely predict that the awakening of a great portion of the Celestial Empire will be mostly due to the civilizing elements dispensed and the influence exercised by Hongkong.... While the Colony progresses and new companies and undertakings are organized, and our external trade increases apace, the veil which overhangs our continental horizon is gradually being rolled up; our increasing enlightenment, which may appropriately be symbolized by the blaze of Electricity, may be taken to be as a beacon which will in due course light up all our neighbors.[52]

This encapsulates the view that mastery of science through technologies such as electricity, especially when given form in a public company, was part of a broader civilizing force. Fraser-Smith argued that companies would demonstrate scientific knowledge and contribute to free trade, social progress, and scientific enlightenment. Science, technology, publicly traded corporate bodies, free trade, and social progress would together roll up "the veil which overhangs our continental horizon," and allow China, too, to benefit from this combination.

HKE's first customer in December 1890 was the government, for street lighting purposes. The expected contract to pump water from the Albany reservoir, which would have helped balance demand throughout the course of a twenty-four-hour day, did not materialize. By mid-1892 the company provided electricity to six hundred bulbs. Four years later, in 1896, the figure had climbed to 3,070, and customers included leading companies such as Jardine, Matheson, Sassoon, A. S. Watson, and Lane Crawford.[53] Not everyone was enamored of Hong Kong's embrace of electric lighting. The permanent Undersecretary of State complained of the cost:

> The annual cost of Lighting the streets will be increased nearly 50 percent by the partial introduction of Electric lighting, and I consider that this large expenditure should not have been incurred without obtaining the previous sanction of the Secretary of State. I am aware that the estimated increase in the Assessed Taxes will more than cover this increased expenditure, but in view of the heavy outlay on other works of local improvement, especially the Water and Drainage works now in hand, it is a question whether it would not have been wiser to postpone the introduction of Electric Lighting for which no urgent necessity has so far as I am aware been shown to exist.[54]

In the electricity affairs of colonial Hong Kong, British government intervention is noteworthy above all for its rarity. The right to supply electricity was in most places contested, sometimes strenuously. This was not the case in late nineteenth-century and early twentieth-century Hong Kong. Both HKE and CLP built their electricity supply businesses until the late 1950s with what appears in a global context to have been remarkably little government oversight. There was during this period no formal franchise and there were no negotiations on electricity prices, with the exception of an agreement in 1929 covering electricity supply to the New Territories that will be discussed below. Electricity in Hong Kong was a pragmatic commercial matter to be dealt with by private companies.

Electricity seems to have been overlooked as a public issue in Hong Kong. Strikingly, the electricity supply companies are generally mentioned in only the most cursory fashion in the Administrative Reports, the annual encyclopedic tabulation of the colony's activities, and the complementary Annual Reports. Robert Fraser-Smith's enthusiastic sentiments found few echoes in the written records of the colony. There was an appreciation for the pragmatic benefits of electricity but surprisingly little sense of the wonderment and the belief that the technology heralded the beginning of a new era for humanity that often accompanied electricity's introduction elsewhere.

Hong Kong capital and expertise helped develop Canton's electricity supply industry. In November 1898, the *Hongkong Weekly Press and China Overland Trade Report* reported the construction of the Canton Electric Light Works Co., "a building in foreign style and nearing completion." The facility was located about a mile downstream from Dutch Folly Fort and the newspaper reported that it was expected to begin operating in early 1899.[55]

Hong Kong had been linked to Canton since the colony's founding in 1841. Indeed, Hong Kong was chosen as a colony precisely because of its position as a protected harbor within easy reach of the southern Chinese city that was China's only substantial foreign trading site.[56] Those ties strengthened over the next six decades. By 1908, steamboat service connected Hong Kong and Canton six days a week, with ships leaving Hong Kong at 9:00 p.m. and arriving early the next morning in Canton; from Canton's foreign district of Shameen (Shamian), the 5:30 p.m. departures arrived in Hong Kong about midnight. The ships were lit throughout with electric light and first-class cabins boasted electric fans, even on the less expensive Chinese-operated lines.[57]

Robert Shewan's firm, Shewan Tomes, had a substantial office in Canton, one that reflected its prominence as an exporter and its status as one of the most important foreign trading houses in China. Wright and Cartwright's 1908 account noted that "the firm deals with the bulk of the articles exported from Canton through Hong Kong." These included, among many other items, goods as diverse as silk, tea, firecrackers, rhubarb, rattan, and palm-leaf fans.

In addition to CLP, Shewan Tomes managed many other businesses, including Hong Kong Rope Manufacturing and the Green Island Cement Factory. The rope company sold its millions of pounds of annual output to customers in Japan, the Straits Settlements, India, and Australia. Shewan Tomes's dozens of agency businesses included finance (Equitable Life

IN THE BEGINNING

Assurance of the United States and the China Provident Fund and Mortgage Co. were two of many financial firms the company represented) and shipping: Shewan Tomes also operated the China and Manila Steamship Co. and its two "first-class boats" on the Manila route.

Shewan Tomes had other electricity companies besides CLP. It "looked after" the Kwangtung Electric Supply Co. and supplied the Sandakan Light & Power Co.[58] In addition to its substantial premises in Canton and Hong Kong, Shewan Tomes had offices in Shanghai, Tientsin, London, Kobe, and New York. It also had agency relationships with firms in Amoy (Xiamen), Foochow (Fuzhou), Formosa, Hankow, Manila, and the Straits Settlements.[59]

Hong Kong's strong links with Canton are evident in the development of electricity in the two cities. Canton's electricity supply plant received important organizational and financial help from Hong Kong. The consulting engineer on the facility was HKE general manager W. H. Wickham. William Danby, of the prominent Hong Kong architecture firm Danby, Leigh & Orange, designed and supervised the construction of the plant. Shareholders, the newspaper noted, were "mostly Canton officials and Hong Kong merchants." The plant included three dynamos and engines from Johnson & Phillips, the supplier to HKE, and three boilers from Babcock & Wilcox. An article in the *Hongkong Weekly Press* reflected the high hopes of the project's backers, who expected to rapidly expand the facility: "The light is to be used for street illuminations in the city, and the applications for private installations have been so numerous that already the duplication of the plant is under consideration."[60]

A key figure in the development of electricity in Canton was Shewan's longtime associate, Shewan Tomes comprador Fung Wa-chuen, whom Cameron credits with setting up the Canton electricity company in 1898. Fung was a graduate of Queen's College, an assistant at the Yan Wo Opium firm on Cleverly Street, and, later, a director of the Tung Wah Hospital, a director and chairman of the Po Leung Kuk, chairman of the Chinese Chamber of Commerce in 1900, and foreign affairs deputy to the Viceroy of Canton in 1909.[61] It may be that Fung, in association with Hong Kong and Canton investors, founded the electricity company, built it with the help of Wickham and Danby, then turned to Shewan for investment shortly after the facility opened, when profits were slower to materialize than expected. This transaction demonstrates the small, entrepreneurial world of Hong Kong traders and the significant ties they enjoyed with Canton.

CHINA LIGHT & POWER: EARLY YEARS IN CANTON AND KOWLOON

The China Light & Power Syndicate was incorporated in Hong Kong April 23, 1900, two years after the start of the lease on the colony's New Territories, with the aim of providing electricity for the undeveloped Kowloon section of the colony on the north edge of the harbor across from Hong Kong Island.[62] Ostensibly founded to light Kowloon, CLP's first significant corporate act was the acquisition in 1900 of the Canton electricity plant, which had started producing electricity in 1899. Acting on behalf of CLP shareholders, Shewan paid $80,000 in cash and $20,000 in CLP shares to acquire the Canton Electric and Fire Extinguishing Co. Lawrence Kadoorie later noted that the acquired company had been established "to provide lighting and fire pumping to Canton."[63] Although Shewan enthusiastically cited prospects for income from the fire pumping operation in the first annual shareholders meeting of the reconstituted CLP in May 1902, no subsequent mention of firefighting appears in CLP annual reports.[64]

The company restructured and raised additional capital in 1901. Elly Kadoorie, Lawrence's father, was among the group of seven investors in the new company, which was incorporated June 25, 1901 with a registered capital of $300,000. Five of the seven shareholders were from Shewan Tomes: Robert Shewan, Anthony Babington, James Duff, Fung Wa-chuen, and Archibald Reid. Elly Kadoorie and Sassoon Benjamin, Kadoorie's partner in the firm of Benjamin, Kelly & Potts, were the two other shareholders. Thus began the Kadoorie family's association with CLP. Each of the seven shareholders was said to have held "one share," but this appears to have been meant in a metaphorical sense as the par value of the shares was $20 each, for a total of 15,000 shares. Original records are not available to confirm this.[65] The CLP investment was not an anomaly since Elly Kadoorie had already exhibited an interest in lighting. The previous year, in 1900, the Kadoorie family had become the largest shareholders of Shanghai Gas Co., founded in 1865 to light Shanghai; Lawrence's brother Horace later served as its chairman.[66]

CLP's electricity-generating station was not the first attempt to set up an electricity operation in Canton, though it may have been the first successful one. Chi Zhang and Thomas Heller quote a Chinese source asserting that the Guangdong provincial government, using imported equipment and local coal, had become the first Chinese electricity producer in 1888. They also note that a British company had established a power plant in Shanghai in 1882. Electricity supply systems designed to provide street lighting and serve wealthy households were established in large cities, including Tianjin, Beijing, and Wuhan.[67] In what appears to be a Canton venture different from the one Zhang and Heller refer to, a Chinese merchant living in San Francisco installed equipment for an electricity-generating facility in Canton in 1890, although there is no record of the project actually operating.[68] The U.S. Westinghouse company was also active in Canton in 1890.[69]

Both Chinese entrepreneurs and government officials were interested in bringing electricity to Canton. In January 1890, Hong Kong readers were told that "some time ago" a Chinese merchant living in San Francisco, Wong Ping-shueng (or Huang Ping-ch'ang), had "applied through the Chinese minister at Washington" for permission to establish an electricity-generating plant in Canton and Fatshan (Foshan). Wong had secured permission and now had "formed a company, purchased machinery, plant & c., and conveyed them to Canton." He had won the support of Chinese officials in the United States as well as the current and previous viceroys in Canton. Viceroy Li Han-chang had issued a proclamation granting approval for the project and urged that "no impediment [should] be placed in the way of the company's constructions." In April, in what was likely the same venture, it was reported that the boiler was too large to be transported through the streets of Canton. A new boiler was to be constructed in Hong Kong, but the last words of a short article in April hinted that there were parties who opposed electricity: "meanwhile the anti-progressive bugbears may be able to throw up mountains of difficulties in their way."[70]

We may never know for certain exactly why CLP bought Canton Electric and Fire Extinguishing. Given Shewan Tomes's prominence in Canton and its involvement in a wide range of businesses, Robert Shewan's demonstrated interest in electricity, and Fung Wa-chuen's involvement in the original Canton venture, it is in no sense extraordinary that CLP bought the Canton operation. The fact that Shewan called his venture China Light & Power, rather than, say, Kowloon Light & Power or Kowloon Electric, and does not include Hong Kong in its name, suggests that Shewan always intended to supply electricity in China.

In 1898, before the Canton plant opened, the *Hongkong Weekly Press* sketched out an idealized version of the Canton electricity supply company's planned distribution network. The wires used for electricity distribution were to be mounted on walls with brackets; one wire would continue on to the foreign enclave of Shameen. The company, observed the newspaper, had received permission from consuls and the Municipal Council to use poles for the wire. The implementation, according to Lawrence Kadoorie, proved challenging: "The distribution system consisted of bare wires hung on China Fir poles easily accessible to those wishing to make illegal connections. These wires were so heavily overloaded as to become almost red hot. In wet weather they gave off steam. Needless to say, units generated far exceeded units paid for."[71]

A combination of financial and operating problems presumably prompted the sale to the China Light & Power Syndicate in 1900 and the subsequent recapitalization in 1901 that brought Elly Kadoorie into the company. Suspicion of and resistance to using electricity by Chinese might have contributed to the Canton operation's poor results. One contemporary observer commented on the suspicion of electricity in Canton: "Before doing business with Eastern peoples it has often been necessary to educate them to appreciate the use of that which the vendors sought to supply. The Chinese, for example did not receive the electric light at all favorably at first, and the China Light and Power Company had much difficulty on that account."[72] Certainly some Chinese, as well as many foreigners resident in China and, indeed, many people around the world, were wary of electricity. The first fan was installed at the Astor Hotel in Tianjin in 1897. Yet the offices of the electricity company itself in Shanghai's International Settlement used punkahs (hand-pulled fans) instead of electric fans as late as 1910. Foreigners in China believed that electric fans led to pneumonia and stomach trouble, though these early worries about electricity were dispelled, and in the Republican China of the 1920s the fear of electricity was less pronounced than in Britain. In Britain, as in the United States, electricity companies in the 1920s and 1930s ran public campaigns to reassure the public about electricity's safety.[73] Lyndon Baines Johnson biographer Robert Caro notes that farmers in Texas were "terrified" of electricity at the time of its introduction in the Hill Country in the 1930s.[74]

In Canton, demand in the first decade of the century grew quickly. By 1908, Arnold Wright and H. A. Cartwright's *Twentieth Century Impressions of Hong Kong, Shanghai, and Other Treaty Ports of China: Their History, People, Commerce, Industries, and Resources* (or *Twentieth Century*

Impressions) noted that the "large power station" provided current to light "all the Government yameens and offices, a great many private houses, and some of the most important streets."[75] In the first six years of operation, "the local prejudice has been overcome and a more modern plant has been installed."[76] Speaking in 1977, Lawrence Kadoorie noted: "In a relatively short time the demand for electric current, especially from the flowerboats [brothels], grew quickly and to such large proportions that the Company's generators were soon being worked at full capacity, with disastrous consequences and frequent blackouts."[77] In Canton, problems initially centered on the difficulty of managing demand, particularly given the unstable technology, rather than on resistance to electricity. Inefficient and unreliable equipment, as well as high coal prices, posed greater difficulties. The need for additional investment, both to meet increased demand and to replace underperforming equipment, would prove to be an ongoing challenge for CLP in its next two decades, both in Canton and in Kowloon. At the 1903 annual general meeting, Chairman Robert Shewan complained that the plant's two electricity-generating sets "had proved quite unsuitable and their extravagant coal consumption made it impossible for us to work at a profit."[78] The company at the time had about two hundred customers in Canton, supplying the electricity for 685 small lamps and a large fire pump.[79]

Politics also intruded. In late 1905 and early 1906, the Canton operation "suffered an entirely new setback," in Lawrence Kadoorie's words, a political boycott in protest against the U.S.'s Chinese Exclusion Act, which restricted Chinese immigration to and residence in America. The act was initially passed in 1882, then renewed in 1892, and made permanent in 1902.[80] Kadoorie recalled: "The Chinese promptly retaliated with a boycott against American goods and other things American. For some reason they believed the Company was American, an impression it took some time to correct. To the long list of difficulties and obstacles, brought about by ignorance and suspicion of electricity, lack of trained personnel, remoteness from plant supply sources, and the natural calamities of fire and flood, was added this extraordinary and wholly unforeseen hazard, a political boycott."[81] Kadoorie's protestations notwithstanding, there was a strong American connection to CLP. Robert Shewan's partner, Charles Tomes, was from Boston, Massachusetts. Shewan Tomes had been founded in 1895 as a successor company to the prominent American trading house Russell &

Company, an American firm that counted Warren Delano, grandfather of U.S. President Franklin Delano Roosevelt, as one of its partners. Shewan had worked at Russell & Company and Fung Wa-chuen had been the comprador.[82]

There were other threads that ensnared CLP in the anti-American boycott. Fung Wa-chuen was, as we have seen, a pre-1900 investor in the precursor Canton electricity supply station, one of seven investors in the 1901 recapitalization exercise that incorporated what we today regard as the original CLP, as well as the comprador of Russell & Co., and its successor Shewan Tomes; he was also a key member of the Chinese merchant elite in Hong Kong and instrumental in blunting the impact of the boycott.

Fung was the chairman of the Chinese Commercial Union in Hong Kong. Government pressure on the merchant elite frightened the colony's business leaders away from supporting the boycott. For his role in frustrating the boycott, Fung was subject to anonymous placards and threatening letters; these prompted his resignation as the Chinese Commercial Union chairman.[83] Boycotting CLP, a company with which a man who was reviled in the Chinese community was closely associated, looks understandable, if not obvious. Kadoorie's dismissal of the boycotters as misguided—"for some reason they believed the Company was American"—appears either ignorant of the facts or naïve, if not deliberately misleading.

The boycott had little long-term impact. Operations soon thrived, despite continuing problems with the generating equipment. (Lawrence Kadoorie later said that the Canton "plant turned out to be worthless.... After repeated trials we gave it up in despair and installed Diesel Engines from Germany. These gave satisfaction and the business in Canton grew rapidly.") The company did well, "too well in fact for the Chinese officials began to repent their concession and to give us trouble by refusing to pay their bills, all from the Viceroy, Chief of Police etc. downwards claiming that current should be supplied to the Mandarine [sic] free."[84]

The financial pressure on the company proved too much. CLP agreed to sell its Canton operations. The event was traumatic, particularly for Shewan, who complained about critics, speculators, and newspaper letter writers. In the end, the company received $1,330,000 for its Canton operations, with the final installment paid on July 31, 1909. It would be seventy years before CLP would sell electricity in China again. The forced sale was a sobering reminder of how an electricity supply company, a capital-intensive, fixed-asset operation, could be vulnerable to larger social and political forces.[85]

Many of the challenges that were to confront CLP in the following decades were already evident in its earliest years. Demands for capital investment never let up; equipment was unreliable; fuel prices were volatile and supplies uncertain—though consumption always proved to be more than expected; it proved difficult to balance the need for adequate generating capacity with a wish to avoid excessive capital spending; and CLP was buffeted by political events.

As far as the electricity supply business was concerned, CLP was in a race for size, both in Canton and later in Hong Kong. It had to compete against in-house electricity generation by large users. To do so, its network had to be large enough to generate sufficient economies of scale to sell electricity less expensive than that which users could themselves produce. To build a sizeable network required significant investment; long-term loans were unavailable, so cash needed to come from shareholders' funds or from operations. Electricity supply companies that priced their current too high would see slower growth in demand, thus curtailing their ability to expand.

As noted in the examples of Calcutta and Melbourne, electricity had been manufactured by individual enterprises before Edison demonstrated the technical feasibility of supplying electricity from a larger-scale central plant. Steam, which was used to power machinery, had long been produced by factories.

Competition from in-house generation of electricity was widespread. The Peak Tram produced its own electricity. Hong Kong and Whampoa Dock Co. noted in a 1903 report to shareholders that it had begun producing its own electricity, which it used to drive three cranes. The Dock Co. was a significant enterprise, one that in 1899 employed 4,510 people and would later be CLP's largest single customer.[86] A newspaper report on the new CLP electricity-generating plant in Kowloon claimed that electricity from the plant, when used to generate steam power, could save half the cost of the water and coal required to drive machinery.[87] If this were true, it is unlikely that the Dock Co. would have built its own generating facility at the same time.

The preoccupation on the part of CLP executives with producing power more cheaply than captive generating sources persisted for many decades. Kadoorie insisted as late as 1959 in testimony to the ESCC that CLP could not possibly be a monopoly because industrial consumers had the ability to set up their own generating units: "The supply companies do not enjoy the privileges of a monopoly and anyone is at liberty to generate his own electricity and, for that matter, to supply it to others."[88]

A larger, more diverse base of customers balanced the demand for current throughout the day, allowing the system to be run closer to full capacity for more hours and thus reducing the overhead costs that each kWh of electricity was required to bear. Newer, larger generating units also produced electricity more cheaply, further strengthening the case for increased scale.

This understanding of the importance of scale in building electricity supply systems was most successfully put into practice by Samuel Insull, a former assistant to Thomas Edison. Insull, who started his business with a single generating station in Chicago, built an electricity supply network that at its peak in the 1920s spanned much of the midwestern United States.[89] The decision to significantly expand CLP's Hok Un power station, made shortly after the Kadoorie family took control of the company, followed a similar strategy. Insull benefited from significant outside financing, including from a robust bond market, something that Hong Kong lacked; Lawrence Kadoorie, although similar to Insull in his quest to expand capacity ahead of demand, struggled to find sufficient capital.

The economics of electricity supply systems are notable in requiring large initial investments in both generating stations and the distribution and transmission network needed to deliver current to customers. The costs of operating the system, especially for labor and fuel, are relatively small in comparison to the fixed costs involved in building the generating facility and transmission and distribution systems. Hong Kong's financial system had a share market, but it had no long-term debt financing. Bank borrowing was limited to fairly small overdraft accounts. There was no bond market and equipment suppliers did not offer credit. CLP needed to fund new purchases using cash flow from operations or by raising fresh capital from investors. In the United States, a deeper pool of funds and a willingness to experiment with new financial approaches allowed Insull to sell bonds to retail and institutional customers, enabling him to use debt financing on a large scale.

Robert Shewan appears to have ignored, or not understood, the need for geographical focus in order to expand scale. Having founded CLP to provide electricity to Kowloon, the syndicate almost immediately bought the Canton operation. Yet the ambition to light Kowloon remained. Around the time of CLP's founding, Shewan bought a piece of land on Chatham Road in Kowloon for a generating plant. Construction in the colony took far longer than expected. At the reconstituted company's first annual meeting, on May 24, 1902, Shewan complained: "The work is dragged along and not being done, as only the Chinese contractor knows how not to do it. Every kind of threat and entreaty has been used to urge on the work but to little purpose, and it's still a case of hope deferred and excuse after excuse for more men not being employed."[90]

Shewan assured shareholders that the plant would be in operation three months later; in fact, generation would not begin until more than eight months had passed. The Kowloon facility began producing electricity February 2, 1903, although almost three more months would elapse before the *Hongkong Weekly Press* wrote that Victoria residents "have hitherto held an advantage over their neighbors in Kowloon in the possession of electric light as an illuminant for their streets, offices, and residences, but the completion of the new lighting station . . . has equalized matters in this direction and placed Kowloonites on a level footing with people on the other side of the harbor."[91] The paper also noted that since its opening the plant had been "running with complete smoothness" and that operations had not been "marred by a hitch of any kind." The company's share price suggested otherwise. Although investors had paid in capital of $20 a share, the company's scrip was quoted at $10 a share throughout the period of February to April 1903, when the Kowloon plant was in its first few months of operation. Shares traded at $8 in late April, after the Kowloon facility had started operations, reflecting investor skepticism about the company's prospects. Rival HKE, like most other publicly traded companies in Hong Kong, traded at a premium to paid-in capital, reflecting prospects for future profitability. CLP shareholders valued the company as worth just forty percent of what they had invested.[92] This was despite the fact that CLP had been capitalized at just one-tenth HKE's value. HKE had raised $3 million with its sale of 300,000 shares at $10 each, while CLP's issue of 15,000 shares at $20 each had netted $300,000.

The following years were difficult. Equipment did not perform as promised and the old generators were, with one exception, scrapped in 1909; newer generating technology, made by Westinghouse, was installed.[93]

Electric light use grew slowly. A proposal to power electric buses went nowhere.[94] Kowloon was not growing as quickly as Shewan had expected. Noel Braga, CLP's long-time company secretary, remembered: "From the beginning, Mr. Shewan had unbounded faith in the future of Kowloon and the mainland. He has visions of a great city arising in Kowloon, with very wide roads stretching out to all parts of the mainland. . . . Few people, however, shared his tremendous optimism, so that one of the China Light Co.'s biggest problems was, and continued for long to be, finance."[95] CLP's financial straits were such that Shewan even considered seeking financial help from the king of Belgium, "who was known to be financially sympathetic to struggling enterprises as the China Light Co. then was."[96]

Histories of lighting typically take for granted the inevitable dominance of electric light. Yet nearly two decades after HKE introduced the colony's first street lights, Wright and Cartwright's encyclopedic tome on Hong Kong opined that gas would continue to be the colony's favored form of street illumination for some time to come: "From the fact that no extension of this system [of electric street lights on Hong Kong Island] has ever been carried out, it may be concluded that the gas lighting, which has been altered to the incandescent system throughout, is regarded as the more suitable form of illumination."[97]

Indeed, gas, not electricity, dominated Hong Kong's public lighting system well into the twentieth century. The gas company, which used Japanese coal supplied by Mitsubishi to produce gas at a large works in West Point, had three thousand consumers when Wright and Cartwright's book was published in 1908; this prompted the editors to proclaim "public lighting is in its hands."[98] (Town gas was manufactured using coal.)[99]

In 1928, almost three decades after HKE had started operations, charges for gas still dominated government lighting expenditures. That year the government spent $94,723.47 on 1,634 gas lamps. That compared to the $42,758.97 spent on operating 687 electric lights and twenty-four traffic control lights, as well as some lights provided by Taikoo Sugar and Taikoo Dockyard. Gas was still used in almost half of new government light installations. During the year forty-six electric lights and forty-four new gas lamps were installed. (These appear to have been for outdoor lighting only.)[100]

The gas company moved into Kowloon early on to provide street lighting; it found little competition. Unlike many electricity supply companies, whose initial focus was on street lighting, CLP concentrated on industrial users. Its street lighting program began only in 1919, and street lighting only became a significant part of CLP's operations in the 1950s.[101]

Electricity is surprisingly absent in official accounts of the colony. The annual Administration Reports provide copious details on the water system, roads, prisons, the postal service, and virtually every other operation of the colonial state. In the case of lighthouses, to take one example, the number of vessels signaled, messages sent, interruptions to telegraphic communications, hours of fog during the year, how many times relief was delayed due to bad weather, and even the number of times each foghorn sounded, are all detailed. In a striking contrast, electricity—run by two private-sector companies—is treated briefly, and only in appendixes.[102] This puzzling omission is most plausibly explained by the lack of government involvement but also must reflect the rapidity with which electricity was absorbed and became an unremarkable part of daily life in the colony.

By the first decade of the twentieth century, Hong Kong's major new buildings were electrified. St. Andrew's Church, which opened in Kowloon in 1906 and was Paul Chater's place of worship, was equipped with electricity.[103] The law courts, which in 1908 were under construction, featured an electric lift for officials and judges.[104] The Cathedral was lit by electric light, though it was punkahs rather than fans that cooled the building until 1923, when electric fans were finally installed.[105]

Even in thriving colonial enclaves, the pace of technological diffusion was uneven. Electricity coexisted with preindustrial ways of living and working. Although Shanghai in the 1920s enjoyed a reputation as a thoroughly up-to-date city, missionary Frederick Dietz's account of his first evening in Shanghai tells a different story:

> Our automobile passed, with continuous coughing of its horn, through ever-narrowing streets, by a multitude of shops most of which were lighted by oil lamps but some to our surprise by electricity—and, curiously enough, these "Mazda" lamps made us feel at home. This was the only modern touch to the picture and certainly jarred with the surroundings. Boys—stripped to the waist, some of them—were still hard at work pounding brass in a shop in front of which we stopped, for the poor Chinese have no eight-hour laws and child-labor prevention.[106]

This vignette from Shanghai, one of the most technologically sophisticated cities in Asia, illustrates the unevenness with which electricity was diffused some four decades after it was first supplied in the city.

Electricity supply companies often struggled financially to survive long enough to achieve the scale needed to reduce electricity rates and thus attract more customers. As noted, CLP itself was restructured in 1901, only a year, after its founding, and again in 1918. Shanghai's original electricity company was reorganized and recapitalized after six years. Swatow's (Shantou) electricity plant closed after a mere four months in operation because of a financial dispute among the directors, with Wright and Cartwright caustically recording that "the Company is entirely managed by Chinese, and the introduction of improvements slow."[107]

British-linked businesses in Hong Kong were relatively untouched by World War I and in fact gained vis-à-vis their German counterparts, who had until the war been a significant part of the colony's European trading community. The Dock Co. benefited from British Standard ships built as part of the war effort and in so doing contributing to electricity demand from what was now CLP's largest customer.[108]

THE KADOORIE FAMILY AND CHINA LIGHT & POWER

The increased demand for electricity during World War I did not generate enough profits to finance CLP's needed new investments. In 1918, as part of a recapitalization plan to raise funds to finance the move from the original CLP site on Chatham Road to a new plot to be built on reclaimed land at Hok Un, CLP was once again dissolved and reorganized, with additional capital raised and assets transferred to a new company, China Light & Power (1918) Ltd.[109]

The Kadoories stepped up their involvement in CLP. Elly Kadoorie's younger brother Ellis invested in CLP for the first time in 1918. Lawrence Kadoorie remembered attending a board meeting with his father at the Shewan Tomes offices in the St. George's Building in 1918, perhaps in connection with this transaction. Kadoorie implied that this visit marked his first association with CLP.[110] Still, the Kadoories do not appear to have been anything other than passive shareholders, neither holding a seat on the board of directors nor involved in day-to-day affairs as managers until 1928.

Elly Kadoorie had come to Hong Kong in 1880, around the time of his fifteenth birthday. Born in Baghdad, he lived for a time in Bombay before arriving in Hong Kong. He initially worked for the Sassoons, who also came from Baghdad and by this time had established themselves as the wealthiest and most powerful Jewish business group in Asia. Kadoorie, who lived in Hong Kong and Shanghai, was similarly successful over time and by the 1920s had retired to London, where he lived as a wealthy man. In late 1926 he acquired British nationality on an urgent basis and returned to Shanghai.

What explains the urgency with which Kadoorie changed from a seemingly carefree traveler, enjoying Europe in the mid-1920s, to an anxious businessman pushing to obtain a certificate of naturalization and return to Shanghai? Kadoorie's motive appears to have been concern about the effect of the widening civil war in China on his business and his desire to be protected as a British subject when he returned to Shanghai. Special Branch inspectors reported, "Owing to the troubles in China he is anxious to return to Shanghai to attend to his business and if possible wishes to receive a Certificate of Naturalization before he leaves," noted the report.

Elly Kadoorie's rushed naturalization process, his abandonment of a leisurely life in Europe to return to Shanghai, and the refocus on Hong Kong in the late 1920s must be seen against the backdrop of the deteriorating political situation in China. In July 1926, Chiang Kai-shek moved north from his base in Canton to wage his Northern Campaign.

Fear of being subject to the whim of warlords, rather than protected by British law, goes a long way to explain Elly Kadoorie's decision to secure British nationality before his return to Shanghai, a move that proved a transformational moment in the history of the Kadoorie family. The documents also provide the key to understanding why the Kadoories renewed their focus on Hong Kong: Elly Kadoorie knew that Hong Kong, a British colony, would likely offer more security if Shanghai's status as an international enclave were threatened.[111] As a newly naturalized British citizen, he would be positioned to make the most of that protection. His urgent acquisition of British citizenship at the beginning of 1927, combined with his return to Shanghai and the expansion of the family's Hong Kong business activities, suggests that he was motivated by the need to safeguard his existing interests in Shanghai and at the same time to nurture new ones in Hong Kong.[112]

Elly Kadoorie's decision to monitor events from Shanghai proved shrewd. In late March an uprising saw the Chinese Communist Party seize much of Shanghai, although not the international settlements. On April 9, 1927, Generalissimo Chiang Kai-shek declared martial law. Chiang began a purge of Communists on April 12, with thousands of Communists and leftists killed and wounded in Shanghai.

As noted, it was only following Kadoorie's return to Asia that the family for the first time took an active role in CLP affairs. In 1928, Elly Kadoorie appears in the CLP corporate archive for the first time since his original investment in 1901, with the announcement to shareholders that he and José Pedro Braga had joined the board of directors.[113] In 1927 Braga, commonly known as J. P., had become the first Portuguese appointed to the Legislative Council. This was at a time when Portuguese were accorded an ambiguous racial status as "locals," situated between Europeans and Chinese. The Kadoories, as Jews, also occupied an ambiguous position.

It is striking how absent the Kadoories are from CLP's records until 1928. As noted, Elly Kadoorie had been one of the original seven shareholders who invested in the recapitalized CLP in 1901. Ellis Kadoorie invested when the company was recapitalized once again in 1918. Yet the Kadoories are found nowhere in the CLP archives before Elly joined the board of directors in 1928. Until that year, CLP had seemingly been peripheral, just one of the Kadoorie family's many investment interests.[114] The Kadoories do not seem to have exercised any management control or had a representative on the CLP board of directors or, as it had been called until 1927, the consulting committee.

The renewed focus on Hong Kong, and the new attention given to CLP, also reflected the increasingly prominent role that Lawrence Kadoorie was taking in the family business. Lawrence had been forced to cut short his studies and return to Shanghai following his mother's death in 1919, at which point he effectively became his father's chief of staff (Lawrence Kadoorie said that he was Elly's "aide-de-camp").[115] Although Elly in 1928 was the first Kadoorie member to join the CLP board of directors, Lawrence played a leading role during the 1930s as part of the charge he had been given to rejuvenate the family's Hong Kong interests. Lawrence's hands-on involvement with CLP was particularly evident in the expansion of the Hok Un generating facility. Lawrence's younger brother Horace was for a time on CLP's board but was less involved in the family's business operations.

In 1929 CLP won the right to electrify the New Territories, a large but sparsely populated part of the colony that had been acquired in 1898 on a ninety-nine-year lease. The same year, the company rebuffed a takeover attempt by its larger cross-harbor rival Hongkong Electric. The Kadoories' involvement with CLP deepened in 1930 when Lawrence Kadoorie joined Elly Kadoorie on CLP's board of directors. Lawrence had moved to Hong Kong in the late 1920s to reopen the family's office there, while his brother and father remained in Shanghai.[116] The family's intention to take full control of the company became apparent in 1931 when directors began negotiations to buy Shewan Tomes out of its perpetual management agreement. In early 1932 the board of directors severed the operating contract Shewan Tomes had held since CLP's founding in 1900. Shewan, who had served as chairman of CLP since its founding in 1901, left the board of directors. The board managed the company from April 1, 1932.

The Kadoories' victory came at a significant financial cost for CLP. An arbitrator in early 1933 awarded Shewan Tomes $2 million, plus costs—double what Shewan Tomes had originally asked in September 1931 and more than eight times what CLP had offered at that time.[117]

The split was an unusual instance of a public rift between the Kadoories and a business partner.[118] It is even more surprising given the four-decade association between Shewan and Elly Kadoorie and Shewan's importance in Hong Kong's small non-Chinese business community; Shewan was a longtime board member of the HSBC, serving on it from 1895 to 1911, including for a time as the chairman.[119] He was also a member of the Legislative Council. The year after Shewan Tomes won the arbitration award, on February 14, 1934, Robert Shewan died after falling out of his window into the garden at his Number Two Conduit Road residence, a building situated immediately below Chater's Marble Hall in Hong Kong's Mid-Levels district. Shewan's suicide, if that is what it was, took place on the fifty-third anniversary of his arrival in Hong Kong.[120]

Shewan's passing symbolized the waning of an era in which powerful merchant trading houses could profitably control a wide range of unrelated businesses. The Kadoories' assertion of ownership rights was part of a transformation that saw greater management specialization, as well as more fixed capital investment, in firms that were of sufficient size and complexity to reward specialization rather than diversification. The treaty-port era, when trading houses like Shewan Tomes operated a slew of largely unrelated businesses, gave way to a new era of increased professional management, engineering sophistication, and capital investment. The Kadoories' takeover of CLP management and the ousting of Shewan Tomes in some sense parallels the ascendance of managerial capitalism that occurred at the end of the nineteenth century in the United States.

The Kadoories remained acutely mindful of the deteriorating political situation in 1930s China. In December 1937, ten months before the Japanese invasion of Guangdong, Horace wrote Lawrence of his fear that the Japanese were stirring up anti-British feeling. Writing from Shanghai, Horace worried that this could affect Hong Kong should the Japanese decide to fight in Canton. Lawrence, who appears the more optimistic of the brothers in these diaries, responded from Hong Kong: "I do not think there will be any trouble here."[121] In the private diary Lawrence and Horace shared, the brothers evince overall optimism and a desire to ensure that the family was positioned for any improvement.

On Lawrence Kadoorie's wedding day in 1938 he sent a document to his father's longtime associate, CLP Chairman J. P. Braga. Kadoorie titled his document "General Policy (A Brief Survey)." This essay is worth examining at length. Kadoorie's decision to send the document on his wedding day indicates its significance. The nine-page essay discussed pricing, capital structure, consumer sentiment, technological excellence, and operational efficiency insofar as they applied to electricity supply companies. Its themes are ones that recur in subsequent decades, reflecting the importance of the principles he laid out for his company.[122]

Kadoorie argued that CLP was not selling a technology or a product but a service—the ability to banish the night or do away with cold: "The product offered to the public is homogeneous—a unit of electricity—a kWh. But its degree of usefulness to the consumer varies considerably. It is essential to regard electricity supply not only as a commodity but also as a service."[123] Kadoorie argued that there was more "obvious human satisfaction" that came from one kWh used for lighting than one employed for heating. One kWh of electricity will light a hundred-watt bulb for ten hours, lighting a "large room on the darkest winter night," while one kWh of heat "will hardly heat the same room for more than twenty minutes."[124]

If electric current was not a commodity but a service, Kadoorie contended, then it could be priced differently for different uses. Consumers should pay more for electricity they used for lighting than for heat, both because it was more useful for them and because they did not have an obvious alternative. Conversely, Kadoorie wrote, CLP was justified in charging less for mechanical power, both because customers had other choices and because their demand was more constant throughout the day and throughout the year, allowing the company to spread fixed capital costs over a greater volume of electricity. Consumers valued lighting but they only used it a few hours in the evening; yet the company had to invest in fixed-cost overheads that needed to be paid for.

Noteworthy was Kadoorie's analysis of electricity as a homogeneous product that could be priced differently according to its final use. When Kadoorie wrote his "General Policy," and until 1969, CLP and HKE had a system of dual meters in order to charge consumers different rates for light and power.[125] Lighting cost more than power; bulk power users benefited from additional preferential rates.

Kadoorie, in short, laid out the rationale for his belief that consumers should pay higher rates and thus subsidize industrial users. The issue of cross-subsidies would become critical two decades later, when industrial and consumer pressure on the company, and on the Hong Kong government, would see CLP face the prospect of rate regulation and even nationalization.

The "General Policy" outlined principles that would underpin the company's growth for the next half-century. Prime among them was the commitment to supply power whenever and wherever customers wanted it, to "be prepared to meet all demands for electricity as and when they are made:"[126] "The economics of electricity supply differ from those of normal industry and business because, once the demand is made for electricity the supply of it must be available to the consumer *always* even though he has no desire for the use of it at the time."[127] The essay addressed an issue that defines the economics of electricity supply systems. Demand fluctuates dramatically during a twenty-four-hour daily cycle, a seven-day week, and over the course of a year. Capital investment must be significant, in order that an electricity supply company would be able to respond to surges in demand for electricity, a product that could not be stored but must be consumed instantly. Kadoorie counseled that capital spending must be forecast at least two years ahead of time so that equipment could be ordered and installed and adequate reserve generating capacity maintained.[128] This was the case even though the company's equipment had to be paid for at the time of delivery.

He advised that the company should "buy the best and most reliable plant, bearing in mind local conditions and the fact that the Company is situated 10,000 miles away from the manufacturers."[129] In short, it should buy good equipment, operate it efficiently, follow technical developments, and prepare for future growth by buying equipment ahead of demand and acquiring land for substations. Property, Kadoorie noted, could be used for "staff quarters or cash offices" until needed. "The board, whilst not buying unnecessarily, has bought of the best and has provided for easy and orderly expansion."[130]

Kadoorie also counseled financial and operational prudence. Bank borrowing should be kept to a minimum: "Only borrow seasonal or temporary Capital from the Bank."[131] This deep-seated aversion to debt would endure, as will be discussed below, until Esso invested in 1964. Until that time, CLP paid for all of its generating and transmission equipment when it was purchased. Its policy differed from that pioneered by Chicago's Samuel Insull, who found that debt could speed up expansion. Insull's projected growth in electricity demand convinced bondholders to accept a rapid increase in debt, with the funds used to buy new generating, transmission, and distribution equipment. The debt could be repaid with the additional revenues that resulted from expansion.

Access to raw materials was another concern. Kadoorie counseled that coal stocks should be kept high, as the cost and quality of coal imports were unstable. Kadoorie also recommended that managers "follow all the latest technical developments."[132] Depreciation should be high, thus cutting reported profits; profits should be distributed to shareholders, rather than kept as reserves. Kadoorie worried that a high return on invested capital would make it more difficult to raise tariffs in an era where dividends were calculated on paid-in capital, rather than, as it is today, on shareholders' equity.

In summary, Kadoorie advised building new capacity in advance of demand; argued that the company's differential pricing was justified because CLP was selling a service, not a commodity; and emphasized that profitability had to be managed in order to retain public support for the company and allow it to set tariffs autonomously. These principles would underpin rapid expansion after 1945. Kadoorie did not explicitly discuss monopoly, but implicit in his discussion of light versus power was the recognition that the company exercised monopoly power in light but not in power. He ended his "General Policy" with a characteristically optimistic assessment of Hong

Kong's future: "As stated earlier it is necessary to take a view either optimistic or pessimistic as to the future. Increased expenditure by the Military has encouraged the Chinese to look at the Colony not only as a Safe Deposit Vault but also as a Sanctuary for China. Increased industry gives good indication of this and seems to justify our taking an optimistic view."[133] Kadoorie's assessment of Hong Kong as a "Sanctuary" as well as a "Safe Deposit Vault" should be seen in light of his father's intense and successful pursuit of British citizenship eleven years earlier. Whatever the troubles in China and Europe, Kadoorie expected that the colony would be a city of refuge.

As late as July 12, 1939, Lawrence wrote Horace that "conditions [in Hong Kong] are full of uncertainty, but so far as I can see, there's no real cause for anxiety. And Hong Kong people are quite optimistic and all the companies are doing well." As for Shanghai, now occupied by the Japanese, "I would advise you not to be too pessimistic. The present times offer a wonderful opportunity to consolidate and put our companies on a sound basis. If this is done we may look forward to the future with confidence and shareholders will receive the full advantage of better times."[134]

Elly, Lawrence, and Horace always worked to ensure that the family was well positioned for better times. At the same time they were careful and calculating, hedging against the worst. Referring to the physical storage of shares that the family owned, Lawrence wrote Horace in the private diary on December 27, 1937, that "it is *inadvisable* to retain these securities in Paris for longer than necessary" because of the threat of war.

THE HOK UN ELECTRICITY-GENERATING STATION

The encyclopedic 1908 volume *Twentieth Century Impressions* has a short section on Kowloon that accords the district the briefest praise. The work notes that "at the present time Kowloon is in its youth, but it is growing vigorously, and gives fair promise for the future, when the Kowloon-Canton Railway shall have linked it up with Peking and the Trans-Siberian Railway." Hung Hom, adjacent to the Hok Un site, is described as "a small village in which the dockhands live." The head of Kowloon Bay is "lined with engineering works, the most important being those of the Hong Kong and Whampoa Dock Co." The book goes on to mention the CLP facility as an afterthought: "There is also an electric light and power station here."[135] The contrast with the CLP facility in Canton, which benefited from a photograph and an extensive writeup, reflects Kowloon's low standing and the modest scale of CLP's operations.

CLP's 1918 reorganization was almost certainly done in order to raise capital for a new and larger generating station to replace the original plant on Chatham Road. In 1919, CLP exchanged the Chatham Road site for Marine Lot Number 93. (Marine lots were ocean "plots" that were sold with reclamation rights.) After reclaiming the site and constructing the new generating plant, the Hok Un facility began producing electricity in 1921. The direct sea access at the Hok Un site allowed the station to have sea water for cooling as well as direct access for coal deliveries.

Hok Un's location between the Hong Kong and Whampoa Dock Co., for many years the company's largest customer, and Green Island Cement, another customer and for many years part of the stable of Shewan Tomes–managed companies, was also beneficial. The Kadoorie family invested in both companies. The symbiotic growth in Kowloon owes much to the interlocking and overlapping investments by Shewan Tomes, Paul Chater, and the Kadoories, as well as to the construction of the Kowloon-Canton Railway. CLP's electricity network, which Kadoorie saw as a "public undertaking," required forward-looking and patient capital, underpinned by "confidence in the future." Kowloon was where the Kadoories placed their confidence. Their investments in CLP's expansion accelerated the rate at which change and growth occurred at the Hong Kong and Whampoa Dock Co., Green Island Cement, and other Kowloon businesses.

Lawrence Kadoorie reminded his audience at the Hok Un opening what Kowloon looked like in the early years of CLP:

> Not even the most lively imagination could have predicted that Banana Avenue with its soft sandy surface was leading to a future city destined to become the terminus of land, sea, and air communications of the first importance in the Far East....
>
> Today the Company's property at Hok Un comprises an area of no less than eight acres. The selection of the new site speaks well for the vision of those responsible. In a public undertaking of this nature, breadth of vision is an invaluable asset; and confidence in the future, backed by the judicious expenditure of required capital ahead of time is essential.[136]

IN THE BEGINNING

The new 12.5 megawatt generating unit at Hok Un increased CLP's capacity almost two-thirds, from 19.5 MW to 32 MW. The plant, which included the world's largest chain grate stoker to fuel the boilers, fit the Kadoories' policy of buying top-flight equipment and investing to meet future demand.

For the 1930s Hok Un expansion, the Kadoories used British consultants Preece, Cardew & Rider, who had been working with the company and with HKE and, indeed, with the Shanghai electricity company, for many years. Seven companies tendered for turbines. The company ordered from Metropolitan-Vickers Export Co. of Manchester. The boiler was ordered from the International Combustion Ltd. of Derby. S. E. Faber, who "had designed the power station in Shanghai and was well known to the Kadoories," was the architect. Kadoorie company Hong Kong Engineering & Construction Co. was the principal civil contractor. The site at Hok Un was built with further expansion in mind, containing space for at least four additional 12.5 MW units.[137]

Lawrence Kadoorie was forty years old when he gave the opening speech at Hok Un. He had spent the better part of a decade planning and overseeing the expansion of the Hok Un site. Lawrence and Horace Kadoorie's private diary shows that Elly Kadoorie was involved with larger decisions, such as changing the siting of some of the buildings, but the day-to-day implementation of the Hok Un expansion was in Lawrence Kadoorie's hands.

Supervising the Hok Un expansion gave Lawrence a chance to develop apart from his Shanghai-based father and constituted the most significant accomplishment to date of his blossoming career as a businessman. He chased down missing, late, and damaged equipment. He negotiated with suppliers and engaged in trouble-shooting. He appears to have approved every payment on the project, initialing invoices with a fountain pen. No detail, however small, seemed unworthy of his attention. A March 1939 letter to Preece, Cardew & Rider refers to a letter Kadoorie wrote on December 14, 1938, and reminded them of a missing air vent tube "and requesting that the Suppliers send us by air mail one to replace the missing article. We do not appear to have heard from you in this connection nor have we received the air vent tube."[138]

When there was a problem, Lawrence Kadoorie ensured that it was solved. He engaged in an extensive correspondence in 1938 with Metropolitan-Vickers concerning the proper techniques for cleaning some of the generating equipment. The British engineers were concerned that the sandblasting method CLP used could damage the equipment. The two parties eventually agreed that cleaning using a wire brush would be acceptable. The extensive correspondence illustrates Kadoorie's negotiating skills and his attention to detail. He demanded and obtained guarantees from Metropolitan-Vickers that CLP would not be responsible for damage to the equipment from the cleaning. The exchanges between CLP and Metropolitan-Vickers underline the unstable nature of the technology, including the lack of clear maintenance protocols.[139]

Another illustration of the technological instability that CLP encountered and one that demonstrated Kadoorie's intimate involvement with the project came with his reaction to the discovery of corrosion in a water pump. On November 20, 1939, a condenser at the Hok Un site was opened to investigate a leak. Severe corrosion was found. On November 23, Kadoorie convened a high-level meeting involving CLP executives and the local representative from supplier Metropolitan-Vickers to coordinate a response in order to "save unnecessary correspondence and discussion when the matter was eventually referred to Messrs. Metropolitan-Vickers in England." It was unrealistic to expect a timely response from the Metropolitan-Vickers headquarters at a time when letters by ship "via Suez" were the fastest ordinary means of communication with Britain. Kadoorie worried that the corrosion could cause leaks in the condenser tubes and thus contaminate the boiler water, which in turn could corrode the boiler and possibly lead to contaminated steam causing corrosion in the turbine itself. Sewage was suspected as a culprit.[140]

By the time of the February 1940 opening ceremony, discussions of missing air tubes and corrosion problems gave way to self-conscious historicizing. What Shewan had started, the Kadoories would carry forward. Kadoorie projected optimism in his opening speech, although a large portion of his remarks was devoted to detailing CLP's early financial difficulties. Perhaps Kadoorie thought a history of past challenges would bolster confidence that the company and the colony would overcome this latest challenge. Implicitly playing on the "barren rock" notion of Hong Kong as a hostile environment that had been engineered into a civilized center of commerce thanks to heroic and largely British efforts over the past century, Kadoorie reminded his audience how dramatic the transformation of Kowloon had been: "What is Salisbury Road today... was a fifteen-foot avenue skirted by two rows of banana trees interspersed with granite pillars surmounted with oil lanterns which shed their dim light on the few passers-by on their way to and from the single deck diminutive steam launch alongside the bamboo pier at Kowloon Point."[141] Kadoorie placed the expansion of the Hok Un plant in the context of Hong Kong's historical progress. At the time of CLP's founding, Kowloon was a place of banana trees and oil lanterns. By 1940, it was a place of progress, development that was made possible by light and power. Governor Northcote, who followed Kadoorie at the podium, highlighted the vision of CLP's founder Robert Shewan in developing Kowloon, a place that Northcote noted had been considered unfit even for growing cabbage when the British took over in 1860 after the Second Opium War:

> What would Kowloon have been today had it not been for the vision and the faith of Mr. R.G. Shewan and his fellow-directors during the first twenty years of this century? Suppose that instead of marching ahead as a standard bearer, the local light and power company had been governed by a cautious policy hesitating twice or thrice before each step forward; would we have had on this peninsula today the amazing development of shops, hotels, and factories, which the last twenty years have seen and which is still, I am glad to say, in rapid progress. I think it very unlikely. The provision of light and power is essentially fundamental to sound progress...
>
> So much for the past: What of the future? The building in which you find yourselves supplies the best answer to that question. For this vast room is designed to house three times the plant which is now within it and which is adequate to all of Kowloon's present needs; and the rest of the station is on the same courageously far-sighted scale. In the face of those facts can any of you doubt that the vision and faith of the present board which sanctioned the huge expenditure necessary for such an enterprise is as clear and as strong as those which animated the Company's founders?[142]

Northcote, underscoring the central role that electricity would play in the continued development of Kowloon, prophesied a great future for this long-neglected part of the colony thanks to CLP's continuing investment in electricity supply. Already Kowloon had seen an "amazing development of shops, hotels, and factories." Electricity and economic growth were coupled. Electricity was "fundamental to sound progress."[143]

The newspapers of February 1940 were filled with coverage of the European war and of Hong Kong's preparations. To build an expensive fixed asset like a power plant at a time of growing conflict posed special challenges—and a great deal of courage, given the very real risk that the assets could be destroyed. The Kadoories had not curtailed the CLP expansion plans. Instead they had doubled down, telling shareholders at the meeting for the fiscal year ended September 30, 1937, that "your directors felt it was prudent to construct at this time a larger powerhouse than was at first contemplated" in order to meet future demand.[144]

In the short term, Hong Kong in some senses benefited from the conflict in China. Refugees increased demands for housing, food, water, and electricity in Hong Kong. CLP's electricity sales grew over 30 percent in 1938, as factories relocated to Hong Kong. China's problems were a challenge for Hong Kong, but they also presented opportunities. This insight a decade earlier had prompted Elly Kadoorie to seek immediate naturalization as a British subject, return quickly to Shanghai, and shortly thereafter dispatch Lawrence to Hong Kong to build a strong alternate base for the family's business interests.

Off the public stage, plans were made for war. During the Hok Un construction project, engineers at Preece, Cardew & Rider, architect S. E. Faber, and CLP managers debated how extensively the facility should be fortified. Plans to build a duplicate control room at the expanded generation station were developed but shelved on grounds of cost. CLP considered the possibility that the government would force the company to build the parallel facility, but the government did not do so. On September 30, 1939, the same month that Germany invaded Poland, CLP wrote supplier A. Reyrolle to instruct Canton Insurance Office Ltd. "to take out war risk under the Government Pool in order that we may obtain the benefit of the lower rate."[145] In December 1939, A. Reyrolle inserted a clause indemnifying the company in case of war. Building CLP's biggest facility to date at a time of looming conflict could prove foolish or heroic. The period until 1945 made it look like a foolish decision. But the expansion proved prophetic. It both anticipated Hong Kong's future and made that future possible.

Electricity was a key technology, one that was widely available in colonies and treaty ports by the end of the nineteenth century. Electricity supply technology, developed in Britain, the United States, and Europe was diffused throughout colonial Asia and contributed to urbanization and modernization. Electricity's introduction simultaneously reflected a desire to be up-to-date and in turn contributed to modernization.

Hong Kong was a regional technological center. Its residents played a central role in the electrification of Canton, an operation that became CLP's first asset. Canton and CLP's early Hong Kong operations showed many of the possibilities inherent in the building of an electricity system. Demand for electricity seemed to have no limit, and the need for new capital and technology in order to meet the demand was just as unrelenting. Although Robert Shewan and Elly and Lawrence Kadoorie spoke and acted as if electricity could be a politically neutral technology, there were repeated attempts to exert political control over electricity in Canton and Hong Kong.

The Canton operations were subject to a political boycott in 1905 and a forced sale in 1909. A stillborn government attempt to control HKE in the 1920s notwithstanding, Hong Kong was an outlier in not regulating the electricity supply companies before World War II. Remarkably, electricity supply systems were barely mentioned in official government records. The extension of electricity supply was noted only in passing in official accounts. Technology in general was widely used but little celebrated.

Technology was diffused by men of empire, men who were not necessarily British-born, or even British subjects, but who functioned within a British sphere, notably a legal system designed to protect property rights and advance British interests.[146] Hong Kong's 1865 Companies Act catalyzed the formation of numerous companies in the colony, many of them backed by Calcutta-born Armenian Paul Chater, Venetian-born Emmanuel Belilios (a Sephardic Jew), and Bombay Parsi Hormusjee Mody, as well as Baghdad-born Elly Kadoorie and his erstwhile employers the Sassoons. These men, along with the expatriate British, German, and American partners and employees in China coast companies, were responsible for the introduction and diffusion of new technologies into Hong Kong.

The Kadoories' assertion of ownership and management control over CLP and the rupture it caused with longtime associate Robert Shewan displayed a determination to advance the family's business interests even at the cost of an important personal relationship. Elly and Lawrence Kadoorie evidently believed that the time when CLP could be run as a sideline by Shewan Tomes was giving way to an era that would demand more professional management and much greater capital investment. The decision to expand the Hok Un generating facility, which was made shortly after the Kadoories took management control, reflected a new way of thinking, one that was articulated in Lawrence Kadoorie's wedding day manifesto. Capital commitments should be made in advance of demand; supply would elicit demand, rather than following increased consumption. Given the lack of financing, existing shareholders could avoid dilution of their ownership stake by using dividend payments to pay for capital calls, thus in effect using operating cash flow to buy more equipment. Generating and distribution technology should be the best available. War would disrupt Lawrence Kadoorie's plans for smooth expansion. But more focused management and a willingness to make large capital investment in high-quality technology would set the pattern for the postwar period.

Chapter Three

WAR, OCCUPATION, AND NEW POSSIBILITIES
1941–1946

On Saturday, December 6, 1941, a fundraising ball was held at the Kadoories' Peninsula Hotel in Kowloon to raise money for the European war effort. Madame Chiang Kai-shek, the Generalissimo's high-profile wife, was among the guests at the gala evening. Ironically or not, for this was a fundraiser, the orchestra was playing "The Best Things in Life Are Free," when the party was halted by an order that all navy and merchant navy seamen report back to their ships.[1] Little more than twenty-four hours later, on Monday morning, immediately after the surprise Japanese attack on the United States at Pearl Harbor, the Japanese Imperial Army invaded Hong Kong.

At the beginning of 1941 British Prime Minister Winston Churchill had conceded that "if Japan goes to war with us there is not the slightest chance of holding Hong Kong or relieving it."[2] Japanese troops had been massing on the colony's border for three years, since they had captured Guangdong in late 1938, highlighting the colony's precarious position. After visiting the colony in early 1941, Ernest Hemingway vividly predicted that its defenders would die "trapped like rats." By the time of the Imperial Army's invasion, well-placed Hong Kong Chinese had for weeks been leaving the colony, traveling to Macao and elsewhere in anticipation of an invasion.[3]

Four days after the Japanese invasion, on December 12, Churchill sought to inspire and hearten Hong Kong's defenders: "You guard a link between the Far East and Europe long famous in world civilisation." Privately, Churchill wrote that he had "no illusions about the fate of Hong Kong under the overwhelming impact of Japanese power. But the finer the British resistance the better for all." British leaders' hopes for drawn-out resistance proved to be wishful thinking. British military planners had expected the colony to hold out for 130 days. It held out for eighteen.[4]

On Christmas Day, 1941, also at the Peninsula Hotel, Governor Mark Young surrendered to Japanese Lieutenant Sakai Takashi. Britain had not lost a colony in battle since its defeat in America in 1781. Historian Philip Snow's contention that the end of British rule in Hong Kong occurred not at the official handover ceremony on June 30, 1997, but in the Peninsula Hotel on December 25, 1941, is hyperbolic, but it captures a sense of the moment's importance.[5]

British surrender was followed by three years and eight months of Japanese occupation, the only administrative interruption in the 156-year British colonial period. The period is sometimes simply glossed over, as in this description of a local businessman's career in a Hong Kong engineering and construction company: "Mr. Ngok has been in the local construction industry since 1937 although, like many others he went back to China in 1941 for almost 4 years."[6]

The wartime occupation of the city has tended to be viewed as a violent and somewhat shameful aberration that did little to curtail Hong Kong's overall trajectory of development. In fact, the war was not an interlude but a violent upheaval that made possible a new order, an episode that set an example for what could go wrong with Britain's colonial rule but simultaneously opened up new possibilities. First came the swift defeat, followed by the ritual humiliation of Europeans. Then came incarceration. Several thousand Allied civilians, including for a time the Kadoories, were imprisoned at Stanley Internment Camp on the south side of Hong Kong Island. Most military prisoners were held at the Sham Shui Po Barracks facility in Kowloon.

Since 1841, colonial Hong Kong had undergone challenges ranging from piracy to plague, from water shortages to strikes. It had never faced an existential threat in the form of an attack by a foreign power. Japanese victories in Hong Kong and in heavily fortified Singapore, billed by Churchill as the "Gibraltar of the East," destroyed the myth of British invincibility and created the conditions for a very different postwar Hong Kong. After

1941 and, certainly, after the Chinese Communist victory in 1949, there was a sense of impermanence, of provisionality, to the British presence in Hong Kong. In the 1950s, the best hope of Governor Alexander Grantham (who served from 1947 to 1957), Lawrence Kadoorie, and others in positions of prominence in Hong Kong was that Britain could retain possession of the colony until the expiration of the New Territories lease in 1997.

The physical destruction caused by the war, followed by the post-1945 influx of Chinese refugees, catalyzed social and political changes and a reconfiguration of Hong Kong's economy. Relations between the colonial state and private businesses were among those that were reconstituted after 1941 and then again after 1945. In the postwar decades, Lawrence Kadoorie and China Light & Power (CLP) played an essential role in this change by organizing and deploying an expanded electricity supply system that created the material conditions for some of the most rapid economic growth the world had ever seen. In turn, three factors—the rapid rebuilding of the city, the expansion of the state's size and reach, and a dramatic increase in population—created the conditions that allowed CLP to expand on a scale that had few, if any, global parallels. CLP's electricity did not create Hong Kong's economic takeoff, but it was an indispensable prerequisite. The electricity shortages that affected most parts of Asia did not occur in Hong Kong.

First, the physical destruction caused by the war required the construction of new homes, commercial properties, and infrastructural facilities, including electricity and water supply systems. Pragmatic government policies, coupled with support by the Hongkong and Shanghai Bank, financed the purchase of a new generator for the Hok Un facility immediately after the war and allowed reconstruction to occur relatively quickly. As a consequence, Hong Kong's built environment was newer than it would have been if peace had prevailed.

Second, the necessity of state involvement in the war and immediately afterward established a pattern that brought the colonial state into activities that had previously been the domain of the private sector. The war effort saw the large-scale mobilization of people and companies for collective action, both before and after the conflict, followed by the imposition of British military rule in the colony for eight months in 1945–1946. The patterns developed during this period inaugurated a new and lasting degree of administrative control, and the extension of the state's reach in a territory that had long prided itself on minimal government.

The expanded state apparatus in Hong Kong was part of a larger global shift. The 1930s depression and the war that followed prompted government planning, mobilization, and coercion on a significant scale. Even in Britain and the United States, two countries that historically ceded primacy to private businesses, the state's role expanded significantly. A salaries tax was put in place in Hong Kong shortly before the war; this was reestablished, in a modified form, after the war.[7] In 1946, rent control was introduced in the colony.[8] This measure would have been seen as heretical in the prewar era but was justified by the exigencies of the physical destruction of much of the housing stock combined with a near-tripling of the population in the first two years of peace, from about 500,000 at the end of the war to 1.8 million at the end of 1947.[9]

Third, refugees fleeing turmoil in China, especially Shanghai entrepreneurs, brought capital, technology, human networks, and a new sense of possibility to Hong Kong. Refugees also provided a source of labor and of consumer demand. By 1961, when the first government census in thirty years was conducted, 52 percent of the population of 3.13 million consisted of immigrants. About half of the 1.64 million immigrants in 1961 had moved to Hong Kong before 1949.[10]

CLP's Hok Un plant had been in operation less than two years at the time of the Japanese invasion. The plant was deliberately damaged by British military engineers, acting with CLP employees, shortly before the British surrender in December 1941. During the years of Japanese occupation it was badly run down. Fuel shortages further curtailed electricity production as the war went on. The facility was restored to full operation after the war. By 1948, electricity sales had surpassed the 1941 peak. In 1955, a decade after the war, CLP's electricity sales measured by kilowatt hours (kWh) were more than quintuple the 1948 figure.[11]

ELECTRICITY DURING THE BATTLE FOR HONG KONG AND JAPANESE OCCUPATION

Electricity use grew dramatically in the twenty-one months between the opening of CLP's expanded Hok Un plant in February 1940 and the Japanese invasion at the end of 1941, due in large part to the colony's increased population. In the fiscal year ending September 30, 1941, the company sold twice as much electricity (57.5 gigawatt hours) as in 1936 (27.8 gigawatt hours), despite a doubling in the price of coal and the imposition of a 10 percent fuel surcharge added to all bills sent out after April 24, 1941.[12] Profits in fiscal year 1941, which ended September 30, were a record $1,241,255.

Then, in December, came the Japanese invasion. Japan's claim to peer status with Western powers in the early twentieth century depended significantly on its ability to harness and adapt nineteenth-century and early and mid–twentieth century technologies, including electrification. This technological prowess allowed Japan to play economic catchup after the 1868 Meiji restoration, and the country represented the first East Asian example of what Alexander Gerschenkron termed the latecomer's advantage.[13] By the early twentieth century, Japan was the only East Asian country that had developed a significant indigenous industrial base.[14]

It also was the only East Asian country to pursue a colonial, imperial policy in the late nineteenth and early twentieth centuries. China ceded Taiwan to Japan in 1895, following Japan's victory in the First Sino-Japanese War. Japan formally annexed Korea in 1910, having effectively controlled the country since 1895. Japan invaded Manchuria in 1931 and established the client state of Manchukuo.

Japanese officials and intellectuals devoted significant resources to thinking about what technology meant and how best to use it for the country. Electricity was a prominent part of this discourse and of technological adoption. As early as the 1920s, Aikawa Haruki had noted that Japan had become a "culture of electricity." In addition to being a Marxist theoretician, Aikawa supported Japan's colonization of East Asia. He believed that large hydropower projects in Manchukuo and Korea would allow these colonies to skip the wasteful use of oil and coal and to bypass the age of steam and carbon. The use of clean, rational hydropower would bring about what Aikawa called the "Asiatic Energy System."[15]

The Japanese turned British technology to their advantage during the invasion of Hong Kong. They used the recently upgraded road to Fanling, near the Chinese border, to speed their move into the colony. "Of course, we British had built a road up to Fanling so that we could get to our golf more quickly, and as soon as the road was all tarmacked and ready, the Japs were off down it," remembered Margery Fortescue, whose husband Tim was a Hong Kong civil service cadet.[16]

The threat of war had, as noted, loomed over CLP during the Hok Un expansion. The plant was one of Hong Kong's most important strategic nodes. It would be an obvious target for capture or destruction in the event of a Japanese attack. As the prospect of war became imminent, the colonial government—or the British military—made plans to destroy the facility in order to deny electricity supply to the Japanese. Some eight or nine months before the Japanese invasion, Lawrence Kadoorie was told "that under certain circumstances I would have to give instructions for the power station to be blown up or otherwise denied to the enemy." Unsurprisingly, Kadoorie was not eager to destroy the electricity supply station whose construction had taken so much of his time for the better part of a decade. "I said I could not give orders to demolish company property without some official authority—this I never got."[17]

When the invasion took place, the plant was deliberately sabotaged. There is surprising variance in the accounts of this important event. Testifying in 1959, Kadoorie claimed that the facility was blown up by Royal Engineers. Kadoorie's statement notwithstanding, a company history states that CLP employees destroyed the plant. "On the eve of the collapse of Hong Kong, one of CLP's engineers, Mr. George Gavriloff, was ordered to blow up the main installations at Hok Un. However, a small part was later salvaged and recommissioned by the Japanese."[18] Historian G. B. Endacott's history of the colony claimed that there was a deliberate and well-executed policy to disable a broad range of strategic installations in Hong Kong: "The electricity generating station, the docks, and all military and other installations of vital importance to the enemy were made unserviceable before withdrawal."[19] None of these accounts are accurate.

Certainly, the scene during the invasion was chaotic. CLP maintenance engineer J. W. Barker, who had arrived with his wife in mid-March 1939 from a position at England's York Corporation Electricity Company, was ordered to report to the Hok Un plant during the first days of the Japanese invasion. Barker tried to drive to Hok Un. Kowloon was being destabilized by triad, or underworld, gangs and other Fifth Column elements working in the service of the Japanese.[20] Barker remembered:

> Many gangs of Chinese were looting and stopping people to rob them. They attempted to stop the car I was driving, but I just drove straight through them and eventually arrived at the station. The police and the Royal Navy were in charge and they refused to let me enter. They were removing important parts of the plant. I found out later that they were unable to carry them

away so they just dumped them in the harbor, from where they were later gathered up by the Japanese with the aid of some Chinese staff.[21]

Engineer Eugene Joffe remembers the plan: "When Japanese were about to take over instructions were given to sabotage the plant." CLP staff stripped the plant of its governors (a key component that regulates the steam flow to the turbine) and waited for Royal Engineers to arrive with explosives to blow up the new 12.5 megawatt (MW) turbine. The Royal Engineers did not show up, so engineer George Gavriloff "collected all available explosives" and detonated them. Most of the machine was undamaged, but the shaft was bent and the Japanese, despite a newspaper photograph to the contrary, never managed to get the unit running—they "camouflaged the damage with wooden pieces," remembered Joffe.[22]

Lawrence Kadoorie does not figure in these accounts. In 1959, however, he told a government commission: "It was indeed a sad day when I last visited the Power Station in December 1941, to direct, as a 'denial' measure ordered by Government, the demolition of our latest turbine, which had just been commissioned."[23]

Staff at HKE's North Point generating station appeared to know nothing of whatever plans might have been made to disable that facility. Reflecting the acute shortage of British fighters, the troops included station manager Vincent Sorby. A veteran of the Boer War, which ended in 1902, Sorby had joined HKE nearly four decades earlier, in 1903. The Hugheseliers, as the unit defending the North Point facility was known, were all expatriates over fifty-five years old who had previously served in the military.[24] Their commander was Major J. J. Paterson from the firm of Jardine, Matheson, who had served in the Imperial Camel Corps in Palestine during World War I.[25] For their part, the Japanese apparently wanted to keep the colony's electricity supply stations intact and refrained from direct attacks on either Hok Un or HKE's North Point generating station.

On the night of December 18–19, ten days after the invasion began, the Japanese landed on Hong Kong Island.[26] They came ashore at three places between Shau Kee Wan and Braemar Point. HKE's station in North Point became the scene of intense fighting not because of the strategic value of the station but because HKE and CLP staff there attacked passing Japanese. During the course of the fighting, the British decided to make the station a focal point of resistance and brought in more troops. It failed, and the station was shut down.

Coates simply states that "an order had come through to close down the power station."[27] While Lawrence Kadoorie had been told "eight or nine months before the Japanese invasion" that he would be forced to destroy the power station, and engineers such as Joffe were aware of the plan, no such preparations appear to have been made at HKE. In Coates's extensive discussion of the fighting at the station in his authorized history of HKE, there is no mention of any plans to disable it. Instead, the HKE staff effected a controlled and orderly shutdown of the plant. Station manager Sorby died of wounds suffered during the fighting.[28] The first CLP board meeting after the war, in September 1945, began with a moment of silence for those who had died.[29] Kadoorie later said that two senior directors, the deputy manager, and several important members of CLP's technical staff died during the conflict.[30] Cameron estimates that about eighty CLP personnel died during the war, including six during the battle for Hong Kong, or as a result of wounds suffered in the fighting.[31]

Electricity to Central, and presumably all of Hong Kong Island, was cut on December 21.[32] Survival required improvisation, and elements of the electricity network were sometimes used in improbable ways. After surrender, those engineers wearing civilian clothes were eventually interned at Stanley. Their engineering skills proved to be transferable from a large electricity supply station to a prisoner of war camp, enable them to jury-rig an electricity supply.

Internees used their engineering and electrical skills to improvise hot plates with smuggled electric wire, using large flat roof tiles as the cooking surface, "to cook food smuggled in over the barbed wire at night," according to A. R. Cox, who worked at HKE from 1924 to 1951. However, the extra load from the hot plates blew the 200-ampere fuses at the Stanley substation. Internees then fashioned "three copper links from a hot-water cylinder and placed them in the fuse-holders at the Stanley substation, replacing the blown fuses. The supply of electricity was then resumed to the camp, and the Japanese were none the wiser."[33]

The search for cooking fuel was an ongoing preoccupation. When the Japanese ran out of fuel to operate the North Point station, the prisoners made "small Chinese chatties [cooking vessels] from old tins" and used "wood taken from floors and elsewhere to replace the hotplates."[34] The

University of Hong Kong was looted for timber and other items of value; the grounds were dug up in search of a rumored cache of coal; damage to the university left the campus looking as if it had been bombed, and the school was said to have suffered the worst destruction sustained by any British university during the war years.[35]

The Governor's Office of the occupying Japanese forces initially took charge of water, gas, electricity, tram, and bus companies.[36] By January 11, street lights were switched back on and electricity, telephone, and water services were restored later that month.[37] An electricity authority, the Denki-sho, was set up to run the colony's electricity supply.

Occupation forces also seized some industrial plants, including the Hong Kong Ice Factory in Causeway Bay. Electricity supply and distribution were turned over to a private company in 1943. This meant the merger of the two electricity companies and the construction of a cross-harbor cable connecting HKE and CLP's generating stations.[38]

Still, electricity supplies remained restricted and apparently never reached their prewar levels. The electricity supply station at Hok Un was put back into service but did not run at full capacity. As noted, the Japanese military forced CLP workers to recover some of the equipment that had been thrown into the harbor: "A number of machines—but not the 12.5 MW set—were repaired, and a restricted supply of electricity was maintained."[39] Surgeon Li Shu-fan records that European experts helped the Japanese repair the electrical system within a few weeks of victory: "Hong Kong once more became the 'Fairy City.' Batteries of glaring light shone from all the Japanese government offices for they wanted to make an impression on the people. It was an illuminating change for the New Order, they said. We were told that although the city lights could be seen for a hundred miles, the Allies would not dare come near enough to bomb Hong Kong."[40]

Li claims that the restoration of electricity and the widespread use of lights had a "cheering effect on the population." The Japanese soon moved to "*enforce* the consumption of electricity in spite of the fact that no further coal would be available after the stock on hand was used up.[41] During the early days of the occupation, the number of bulbs to be used and the wattage for each bulb were prescribed and a deposit for every existing electric outlet was demanded by the Formosan-Japanese syndicate that controlled the electricity supply. Li's hospital needed electricity for its X-ray machine, especially necessary during the war because X-rays could locate bullets and shell fragments. The number of lights in the hospital, however, meant that a deposit of over 5,000 yen or about $3,500 was needed: "We knew it was daylight robbery and that we would never see the money again."[42] Consumers had to pay a deposit of $5 for the first five bulbs they installed; additional bulbs required a deposit of $2.50 each. Kwok contends that as the occupation wore on the Japanese wanted to depress electricity consumption: "But while the Japanese occupation forces squeezed the population mercilessly, they found this was not enough to curb the use of electricity."[43]

More stringent measures were adopted as the war continued. A fuel shortage forced power cuts in 1944 and 1945.[44] One-third of the colony's cinemas were closed. Later in the war, cinemas were restricted to one showing at 7:30 p.m. each evening. Four cinemas—Meiji, King's, Majestic, and Alhambra—had additional showings of Japanese movies on Sundays and holidays aimed at the Japanese military audience. Restaurants had to close by 11:00 p.m. "and had to resort to kerosene lamps. Next to be cut were elevators and pumps and all lights had to be out by midnight."[45] A Fairy City was reverting to a pre-electrical past, with, as Kwok records the transformation, kerosene lamps replacing those that had been powered by electricity.

Even with the new 12.5 MW unit at Hok Un out of service, the plant was substantial. The Japanese controlled an electricity-generating facility serving Kowloon and the New Territories with a rated capacity of 19.5 MW, comprising the existing facility at Hok Un. Given that the maximum demand CLP had ever experienced was 16.2 MW and that the population had dropped by about two-thirds during Japanese occupation, the problem was not lack of generating capacity.[46]

The Japanese did a poor job of supplying electricity to Hong Kong. During their occupation they were not even able to restore electricity supply to prewar levels, let alone surpass them. They appear to have suffered to a degree from technical difficulties in operating the plants and from the deliberate damage that was done to CLP's Hok Un electricity supply station during the invasion. But the biggest constraint appears to have been the supply of coal. Electricity supplies were progressively curtailed until they were almost nonexistent by 1944. This shortage of fuel echoed the British experience. Access to high-quality, reasonably priced coal had been a chronic concern of the managers of Hong Kong's electricity-generating stations in the half-century since electricity had been introduced in Hong Kong.[47]

American air raids in the summer of 1944 destroyed Hong Kong's oil supplies and disrupted new shipments, diminishing the supply of electricity throughout the territory.[48] Diaries from those held at the Stanley Internment Camp repeatedly mention electricity and fuel shortages in 1944 and 1945. By June 1944, internee Raymond Jones noted that hot plates were disconnected "as the amount of elec. now allowed will not permit their use . . . Use of electricity in HK greatly curtailed. Slowly the public & ourselves are being deprived of essentials" (diary entry June 11, 1944). On August 20 that year he noted: "Electricity off indefinitely (for perhaps 4 days) so got our wood boiler going."[49] Also in mid-August 1944, Barbara Anslow noted in her diary: "Electricity went off for good (we were told)."[50] The supply of coal was so uncertain that the electricity needed to operate Dairy Farm's refrigerators could no longer be counted on. Prisoners on what otherwise approximated a starvation diet were fed partridge and pheasant for fear that it would spoil. "NO ELECTRICITY. No bread. Pheasants from cold storage," read a diary entry.[51] The following day Jones noted that despite no electricity because of a lack of coal, the expected shipment being delayed by bad weather, internees "enjoyed our pheasant rissole, first meat for eight months."[52]

Near the end of the war, as coal supplies ran short, the Japanese burned rice husks and pine trees to fuel CLP's Hok Un boilers.[53] Blackouts were common. *South China Morning Post* editor Harry Ching's diary in July 1944 reported on the electricity situation: "Next door Japs fined ¥100 and to attend gendarmerie three days for penance for breaching black-out. We no juice so every night black-out for us."[54] Hong Kong's electricity stations had originally relied on Japanese coal, although the price, quality, and perhaps also politically motivated disruptions had prompted them to seek other sources of supply.

Japan's inability to supply Hong Kong's electricity-generating stations with coal, given that the country itself possessed significant coal resources, is noteworthy and historically ironic: In 1905, the Russian fleet's desire to economize on coal consumption had prompted it to take the shorter route to Vladivostok, which led the fleet through the Tsushima Strait, where it was destroyed by the Japanese navy. Japan's thoroughgoing victory in turn gave the country the confidence to believe that it could hold its own against any enemy. This confidence contributed to the development of a more nationalistic, militarized culture that led to its calculated decision to attack the United States in 1941: "Thus the attempt to achieve ultimate economic security plunged Japan into the ultimate economic disaster."[55]

As the war drew to a close, Hong Kong reverted almost to a preindustrial age. Colonial Secretary Franklin Gimson, head of the British detainees at Stanley, was among those deputed to carry 250 tons of wood uphill to the camp kitchens to be sawed and cut in late July 1945.[56] Even after the resumption of British control, and CLP management returned to the plant, wood was the basic fuel source for some time. In September 1945, Lawrence Kadoorie marveled at the 400 tons of wood a day that women and a few men were pushing into the giant Hok Un boilers.[57]

THE BRITISH RETURN

It was not a foregone conclusion that Britain would rule Hong Kong after the war. The combination of the 1941 Japanese victory in Hong Kong, a 1943 treaty signed by Britain and the Kuomintang (KMT) formally ending the treaty port system, and debate about whether British or Chinese forces would occupy the territory after Japanese defeat in 1945 added to uncertainty about the possibility of renewed British rule.

American ambivalence regarding Britain's colonies added to the uncertainty. President Franklin Delano Roosevelt had little patience with the continuation of European colonialism and wanted to see Hong Kong revert to Chinese rule after the end of the war. At the November 1943 Cairo Conference, Roosevelt promised Chiang Kai-shek that the United States would help China recover Hong Kong. Chiang, commander of the Allied theater in China, instructed Wellington Koo, China's ambassador to Britain, to negotiate Britain's relinquishment of Hong Kong, or at least the New Territories, with its ninety-nine-year lease. Britain's Foreign Office "suggested that Hong Kong was already a lost cause, so a gesture of cession to the Americans would demonstrate that the British were not fighting the war for the reactionary purpose of preserving the British empire."[58] After an internal British government debate, with the Colonial Office urging that Britain reestablish postwar control of Hong Kong, Churchill supported attempts to resume sovereignty in Hong Kong after the war. In 1943, a Hong

Kong Planning Unit was set up in London, partly to consider how Britain might be able to take physical possession of Hong Kong when the conflict ended. Plans for political reform were developed to forestall pressure to surrender the colony.[59]

Although it was significant that Roosevelt's successor Harry Truman, who took office in April 1945, supported Britain's continued control of Hong Kong, it was China's civil war that proved decisive in enabling Britain to resume sovereignty in Hong Kong. Communist guerilla forces were in Hong Kong because of their resistance work in the New Territories. KMT troops were on the march and potentially in a position to move into Hong Kong. Only the fact that neither the KMT nor the Chinese Communist Party (CCP) wanted the other to take the territory allowed British Admiral Cecil Harcourt to enter Hong Kong and fill the vacuum left by the Japanese defeat.[60]

Harcourt arrived in Hong Kong with a Royal Navy task force on August 30, 1945. The following day he declared a period of military administration, one that was to last for eight months. Brigadier David MacDougall and his team at the Hong Kong Planning Unit took over administration of the colony after they arrived on September 7.[61] Harcourt, nominally acting on behalf of Supreme Allied Commander in China Chiang Kai-shek, formally accepted Japanese surrender on September 16. MacDougall and other administrators then put in place the conditions for rebuilding Hong Kong. The template for a larger colonial administrative state in Hong Kong was provided by the experience of the military administration that ruled the colony for eight months. There was both change and continuity. MacDougall had been a cadet in prewar Hong Kong and had made a dramatic escape by sea on December 25, 1941, with Admiral Chan Chak.[62]

Price controls were imposed in 1945 for a variety of goods and services, ranging from pork and rice to daily wages for unskilled laborers. The controls showed the increased role of the state, even in traditionally laissez-faire Hong Kong, in the postwar world. Rice controls were particularly significant. The government bought about 60 percent of imported rice and sold it through a single wholesaler. Rice rationing cards were issued, entitling holders to about eight pounds of rice every ten days in the 1945–1947 period. These cards also played a role in the distribution of other foods and basic consumer goods, including edible oil and preserved milk, sugar, and soap. Quilts, blankets, and knitting wool were issued irregularly to rice-card holders. Separately, firewood tickets were distributed. Most controls were removed in 1947 but rice controls persisted until 1954, and import licenses were required for rice even after this date.[63] Imports of coal were also regulated until 1955. By that time, coal use had fallen from the prewar level of 50,000 tons a month to 8,000–10,000 tons, as the electricity supply companies as well as more ships, trains, and factories switched to oil for power.[64]

A salaries and profits tax was introduced in 1947. The tax, something like an income tax but that left capital gains from stocks or property untouched, was raised from 10 percent to 12.5 percent in fiscal 1951. The tax was imposed on company profits, individuals' salaries, interest, and income from property. Petroleum was among one of the small number of commodities on which import duties were imposed, and the electricity supply companies were the largest oil importers.[65] Other government initiatives were adopted to improve economic outcomes. In October 1945, the Fish Marketing Organization was established to control the transport and wholesale marketing of fish in order to cut out a middleman (*laan*) system that had kept fishermen indebted and poor. In 1953, the Vegetable Marketing Organization was set up to make loans to farmers.[66]

MacDougall worked largely independently of London, and his administration reached "a level of efficiency that made it the most shining example of all the territories liberated from the Japanese," in the words of historian Steve Tsang. "Banks and utilities were given assistance to enable them to resume operation as soon as possible."[67] Noteworthy was HSBC's decision to finance the order of a new electricity-generating unit, apparently at a time when Kadoorie was still incommunicado in Shanghai and decades before the bank generally provided this sort of financing. Although the circumstances of this transaction are unclear, it is unlikely that the bank would have risked financing a piece of equipment whose value was almost equal to one year's profits without government support.

Administrators had to deal with strong population growth and calls for political and social reforms even as they rebuilt the city from ruins. The colony's population went from about 500,000 when the British returned to well over one million by early 1946.[68] "It was not a period of muddling through by professional soldiers who knew nothing of good governance. On the contrary, it was a period when the administration functioned remarkably efficiently in very difficult conditions."[69] There was a recognition that the prewar political and social system would have to change in some way. Plans for change had been mooted among officials interned at Stanley, including Hong Kong civil service cadet Tim Fortescue, who had used his confinement "to prepare for a new and possibly different postwar administration."[70]

Governor Mark Young returned at the end of April and restored civilian rule on May 1, 1946. Harcourt, MacDougall, and John Keswick, the Jardine, Matheson taipan who had been a wartime adviser at the British Embassy in Chungking (Chongqing), thought that the government needed to allay anti-European feeling. Snow cites the notable lack of enthusiasm for the official British return to Hong Kong on August 30, 1947—an indifference that bordered on antipathy—as evidence that the British resumption of sovereignty had little support. Snow contends there were four times as many of the Republic of China's white-sun flags as British flags after the Japanese surrender.[71]

Young put forward proposals for political reform that would have provided for direct elections to a powerful Municipal Council. His successor, Alexander Grantham, who took over from Young in 1947, deemed the proposals unnecessary, since he believed Hong Kong would never be a self-governing entity. The Young proposals, which faced significant opposition from the business community, were stillborn.

The political reforms proposed by Young, and the "spirit of '46," which envisioned broad changes, helped nurture new forms of political and social activism.[72] The post-1945 period was one of rapid change. In 1946, the year that opium was finally outlawed in Hong Kong, CLP endured the only significant strike in its history. In 1948, British attempts to drive out squatters through the forcible demolition of housing in the Kowloon Walled City showed the limits of British state power and authority.

Resistance and internment during the war proved important in eroding racial barriers. The historian Philip Snow notes that despite the suspicions among the British, Nationalists, and Communists, each with their own agenda, "joint resistance also started to breed between the British and their Asian helpers a new camaraderie unimaginable in the prewar years."[73] Even as conservative a figure as Grantham, who had served as a colonial official in Hong Kong in the 1920s and 1930s before returning as governor, noticed that the prewar insularity and snobbishness had broken down. Grantham noted "a greater mixing of the races" and proclaimed that "the age of the 'blimps' is over" before conceding that "a few of them still remain, even in Hong Kong."[74]

The outright snobbery and racial prejudice that had characterized prewar British rule continued but was no longer tolerated to the same degree. Still, the prestigious Hong Kong Club, along with many other clubs, did not admit its first "local" members until 1964.[75] The "local" designation comprised Eurasians, including Macanese Portuguese and Chinese. The Kadoories had never been subject to this rule against "local" members; Elly Kadoorie joined the Hong Kong Club in 1905.[76] Ellis Kadoorie and Albert Raymond, who was later CLP chairman, were noted as club members in 1915, when Vaudine England says they "ran the Hong Kong Stock Exchange."[77] Lawrence joined in 1932.[78]

RESTARTING THE ELECTRIC CITY: KADOORIE AFTER THE WAR

Lawrence Kadoorie heard about the nuclear bombs dropped on Hiroshima and Nagasaki and the Japanese surrender when a neighbor who had listened to a shortwave radio broadcast furtively told him the news at the family's Marble Hall mansion in Shanghai.[79] After the fighting ended, Kadoorie managed to get aboard one of the first Royal Air Force planes from Shanghai to Kunming and then, registered as "baggage," because of a prohibition against civilian passengers, flew on to Hong Kong.

By his own account, Kadoorie was the first British civilian to arrive back in the colony from Shanghai.[80] He landed two weeks after Admiral Harcourt's fleet and three days before the formal Japanese surrender, and he went almost immediately to the Hok Un electricity supply station where British troops and CLP employees had already restarted the plant. "The main thing was to get busy again. And without electricity you couldn't get busy," Kadoorie later remembered.[81] A telex dated September 17 (four days after he arrived back in the colony) and headed "Press Release" quoted Kadoorie as expressing "unlimited optimism" regarding Hong Kong's reconstruction and said that he looked forward with great confidence to Hong Kong's future as a British colony now that a sound currency had been established: "War damage nowhere near scale expected."[82]

On September 28, 1945, Kadoorie took the time to compose a lengthy letter to a "Mr. Stevens" (Kadoorie's spelling, hereafter corrected to Stephens), a director of equipment supplier A. Reyrolle & Co. Ltd., the Tyneside engineering firm that had supplied switchgear to the Hok Un electricity-generating station: "I arrived here on September 13, 1945, from Shanghai, and am now taking this first opportunity to send you a few lines. General [Eric] Hayes, G. O. C. [General Officer Commanding] all British Troops in China, very kindly sent me down by the first R. A. F. plane to leave that City, with instructions to report to Brigadier [David] MacDougall to help in the reconstruction of the Colony. My wife and family are still in Shanghai."[83]

Preexisting human and technical networks were already being rebuilt in those first few weeks. Kadoorie had already recreated at least part of his corporate infrastructure, given that this letter bears the notation "lk;mab" at the bottom of the last page, apparently referring to the assistant who typed the letter, likely from his dictation.

The electricity supply network, too, was well on its way to being rebuilt. Kadoorie's letter describes the surprisingly light structural damage to the facility, a wartime bomb notwithstanding. The picture Kadoorie paints is a mix of early industrial-era techniques, of women with babies on their backs feeding firewood into boilers, with modern technology, including a large boiler, an underwater high-voltage electrical cable, and steel.

> The Power Station at Hok-Un [sic] is in quite good shape, though a bomb dropped through the roof and burst near the foundation of No. 8 Turbine. . . . Faber who designed the Station and is now here, thinks that there will be no difficulty whatsoever in putting it right later on and that little or no new steelwork will be required.
>
> The boiler House appears to be in good condition. It is an extraordinary sight to see the 200,000-lb. Boiler Unit being operated on firewood. A large number of women (some with babies tied to their backs) and a few men are busy continuously pushing this wood into the boiler.
>
> At one time, shortly after the surrender, when the China Light & Power Co. were feeding Hong Kong through a land cable jointed in nine places, which had been laid under the harbor, we were burning as much as 400 tons of wood a day. Quite naturally the cable soon broke down, consequently lessening the load. We now burn about 140 tons a day. Thanks to the superb efforts of our Staff, the China Light & Power Co. maintained operation ever since the surrender, which is most important as light and power are badly needed by the Authorities. As yet, the Hong Kong Electric is not in operation.
>
> The distribution System is, as far as we can see, in fair condition. Several transformers and switch tanks were punctured and the oil stolen, probably

for cooking purposes. By the way, [CLP Manager D. W.] Munton, when in Stanley Camp, used some transformer oil for frying rice. He states the results were not too bad, though he prefers butter.[84]

The remarks on cooking with transformer oil allude to the shortages of basic foodstuffs during the war, but they also demonstrate that Kadoorie retained a sense of humor. Kadoorie does not mention the role played by the Royal Navy. Official CLP accounts and statements by Kadoorie in the fifteen years that followed the war, notably in testifying at the ESCC, overstate the war damage and at the same time understate the role played by the British military in reestablishing a secure supply of electrical current. More recognition of the British military's role has been given in recent years. Michael Kadoorie in 2008 acknowledged the military help.[85] So, too, did a 2017 documentary: "With power from a British submarine, China Light & Power limped back into action."[86]

Curiously, a table detailing CLP's electricity output starts on August 29, 1945, one day before Harcourt's fleet and, presumably, the British navy, arrived. The output that day was 61,300 kWh, a little more than 50 percent above the daily average of just under 40,000 kWh for the period ending November 1, 1945.[87] This start date suggests that CLP staff who had been prisoners of war or civilian detainees, as well as those who were not detained, went to Hok Un almost as soon as they were released to take over or restart the generating station. It also confirms the fact that the station had suffered minimal damage and could soon be back in operation.

Because of their importance in postwar reconstruction, the electricity supply companies were subject to military administration for six weeks longer than the rest of the colony, until June 15, 1946. British military engineers, with CLP staff, ran operations during these nine and a half months. At CLP, a Flight Lieutenant Malloy signed documents as "Commander of the China Light & Power Co., Ltd." The military control over Hok Un's operations, a period of corporate-state symbiosis, has been underresearched and underanalyzed. In some cases, such as with Kadoorie's later testimony at the 1959 ESCC hearings, this may have been an attempt to shade the historical record. Kadoorie's 1959 remarks emphasized the physical destruction of the electricity supply system, promoted the idea that CLP was the product of pure and unalloyed private enterprise, and made no mention of any military support in rebuilding the generating and distribution facilities. His statements to the ESCC are part of an attempt to rewrite history and claim that CLP had succeeded despite destruction of its facilities by the colonial state itself and that it had rebuilt a severely damaged facility after the war solely through its own efforts: "Government made us no payment for compensation or rehabilitation."[88]

The exigencies of the immediate postwar period nurtured closer cooperation between CLP and various parts of the colonial state than had been apparent in the prewar period. To guard against a possible KMT invasion, the New Territories was heavily fortified, in parts looking like "an armed camp—a situation which, apart from its other consequences, meant that China Light was under pressure to extend its supplies as and when required."[89] During the post-1945 period, some 100,000 KMT troops passed through Hong Kong en route to north and northeast China, a visitation that Kadoorie remembered as being "like a plague of locusts. Houses left intact by the Japanese, who had prevented looting, were stripped bare in 24 hours."[90] Significant increases in the price of coal and other imports led to electricity rate hikes in October 1945 and again in June 1946, when the company reassumed management control from the military. The government was involved in an unprecedented fashion with setting electricity prices. Private companies and the government tried to impose some measure of predictability and even fairness in a war-induced inflationary time.[91] This was the basis for Hong Kong's first price controls on electricity rates.

The June 1946 rate increase was set at 3.6 times the prewar level in detailed negotiations between Kadoorie, on behalf of CLP, along with representatives from HKE and the government, including MacDougall and, behind the scenes, the governor.[92] As part of the agreement, the electricity supply companies agreed that rates would not be raised further without government permission, that the cost of renting an electricity meter would be frozen at $2 a month, and that electricity tariffs should be reduced "as soon as there is any appreciable reduction in the price of coal."[93] At the time, CLP was using 22,000 tons of coal and 19,000 tons of fuel oil annually.[94] Charges were set at 71.28 cents per unit (one kWh) of light; 27.72 cents per unit of power. Kadoorie had wanted the rate set at four times the prewar level and engaged in a good deal of correspondence on the subject. During the ESCC hearings he appeared to recollect almost nothing of these negotiations.[95] It is possible that Kadoorie had simply forgotten, but it is likely that he did not want to discuss the ongoing government involvement in electricity pricing that began immediately after the war.

Fluid human networks also proved important in quickly increasing the electricity supply. Germans and Austrians had been at liberty during the war, among them Jews who had escaped Nazism. Paul Steinschneider, an Austrian Jew who had come to Hong Kong after fleeing the Nazis, had initially been employed by the Hong Kong and Whampoa Dock Co. before joining CLP.[96] Because Steinschneider was Austrian, the Japanese had allowed him to keep working at the plant. After the war he knew how the Japanese had modified the system, particularly the distribution system, and was able to tell returning staff how the system had been jury-rigged.[97] Edgar Laufer, a German Jewish refugee who had fled the Nazis and been employed by Lawrence Kadoorie at CLP in the late 1930s, was able to move freely through the city; he visited different internment camps, bringing families news and sometimes food. It is unclear whether Laufer, who at the time of his retirement was said to be CLP's longest-serving employee, remained employed at CLP during the war years.[98] After the British return, some three hundred to four hundred Japanese helped in the rehabilitation of the electricity supply plant. "The Japanese were so eager to please that they'd pick up anything with their bare hands," remembered former CLP chief engineer George Gavriloff.[99]

This government involvement in electricity supply did not stop in 1946. The colonial secretary met the electricity supply companies' chairmen on November 19, 1948, and suggested that they should reduce electricity rates and restrict dividends.[100] This attempted government intervention was a sharp departure from prewar government policy. CLP responded by seeking legal advice. Given that there was no formal franchise or enabling legislation, the legal advice CLP received was that the government would have a very difficult time pressing its case. In the end, after protracted and intense discussions that also involved the financial secretary, the government backed away from formal intervention. Nonetheless, it was yet another demonstration of the ambitions of some of the colony's most important officials to play a more active role in the economy. These government interventions in electricity were part of a larger pattern of administrative involvement. They helped establish the context for the ESCC's hearings in 1959 and the commission's recommendation that the electricity supply companies be merged and nationalized.

For all the provisionality and fragility of the post-1945 period, new possibilities emerged. In 1956, building regulations were relaxed to allow buildings up to twenty stories high.[101] Taller buildings had more floor space and used more electricity, requiring further expansion of the colony's electricity systems.

The colony's increased population and the rapid addition of new electricity supply in turn allowed for a new scale of economic growth. To make this possible, the Hok Un generating plant needed to be both repaired and expanded. CLP reached back to expansion plans developed before the war. Buoyed by strong growth after the 1940 expansion, CLP had ordered a 20 MW generating turbine. This represented almost an additional two-thirds expansion of its Hok Un facility (from 32 MW to 52 MW) and was more than the entire 19.5 MW generating capacity before the 1940 expansion. The 20 MW unit was ready for shipment when the Japanese invaded Hong Kong. The order was taken over by the British government and the unit was shipped to wartime Russia.[102] Immediately after the war, plans to secure a new turbine were put in place.[103]

The company's nearly half-century relationship with Metropolitan-Vickers, for whom Shewan Tomes had acted as the Hong Kong agent, made CLP a preferred purchaser of the company's electricity-generating equipment. Thanks to that relationship, and a quick decision on financing from the bank, CLP was able to obtain a turbine that had been destined for South Africa.[104] The financing had to be finalized before Kadoorie had returned to Hong Kong, or at any rate before he could be contacted. Frank King tells the story from the bank's perspective: "Payment had to be guaranteed and there was no officer of the utility companies available; once the equipment had been approved as part of Hong Kong's immediate rehabilitation program by the Hong Kong Planning Unit, however, the guarantee was provided by the Bank under the signature of Arthur Morse. Hong Kong thus 'jumped the queue' for postwar orders and the basis for the Colony's recovery was established."[105]

King notes that the cost of the turbine and boiler was equivalent to roughly one year's prewar profits for the bank, so the decision to proceed without Kadoorie's go-ahead was a risky one. Important as the decision to finance the generating unit was in the colony's postwar reconstruction, there were points of friction between the government and CLP.

Some of the plans made in London while Kadoorie was interned or otherwise out of contact did not meet with his approval once he was back in the colony.[106] On returning to Hong Kong and finding that damage to the plant was less than anticipated, he canceled plans for an 11,000-volt switchgear unit, writing to Stephens: "From what I have written above, you will see that the plan made in London, which was based on the Colony being practically wiped out, will have to be considerably altered."[107] Reconstruction nonetheless moved ahead quickly.

Kadoorie wrote Stephens that he was optimistic in part because of his confidence that he could sell electricity to China: "Though Hong Kong has suffered a severe blow, I am optimistic as to the future, especially as I feel there is a reasonable chance that we shall be able to supply electricity to Canton and other points on the Chinese Mainland."[108] Kadoorie kept the China dream alive for a while longer. In late 1945, he negotiated with Generalissimo Chiang Kai-shek's brother-in-law T. V. Soong and other KMT officials to sell electricity to Guangdong. It "could be arranged to supply 500 KW at the frontier and this could serve as a useful model." Negotiations were halted after Kadoorie balked at requests to pay "cumshaw" (bribes) to officials.[109]

When the attempts to sell electricity to China were stillborn yet again, Kadoorie instead brought Chinese industrialists to Hong Kong in the years after the war ended in 1945, but particularly after the Chinese Communist victory in 1949. By persuading Shanghai textile entrepreneurs to move to Hong Kong, he laid the foundations for CLP's rapid growth and for Hong Kong's manufacturing success in the following three decades. His longstanding desire had been to put "China" back into CLP for the first time since 1909—or more accurately, to put CLP back into China. The 1949 revolution would force him to defer that dream for three more decades, although even during China's darkest moments he clung to the belief that he would someday sell current to the PRC.[110] While he waited, he re-created Shanghai in Hong Kong. CLP's electricity provided the underpinnings for the territory's transformation from a slow-paced colony into one that not only overtook Shanghai but also became East Asia's most dynamic, outward-looking, and economically vibrant city.

Chapter Four

"A PROBLEM OF PEOPLE"

1947–1958

The total mobilization during the war laid the groundwork for large-scale, state-directed action for postwar construction in Hong Kong, as it did in many other places. The reconstruction efforts after 1945 occurred in the global context of an intense economic, ideological, and political struggle between a largely Anglo-American-led market-oriented democratic capitalism and a Soviet- and, later, Chinese-led Communist alternative.

For colonial Hong Kong, the 1949 Communist victory in China was a worldwide event with local impact. Hong Kong felt the ramifications of the revolution in the arrival in the colony of more than one million refugees, and in the impetus to action that the expanded population gave to the colonial administration and to business leaders like Kadoorie. The influx of people fueled expansion of housing, medical, transportation, communications, and electricity systems.

In Hong Kong, where a strain of militant anticommunism coexisted with experiments in housing, education, finance, and labor, government interventions reflected political changes in Britain. There, Churchill's defeat in the May 1945 elections was followed by a six-year Labour government under Clement Atlee, which, emboldened by the 1942 Beveridge Report (*Social Insurance and Allied Services*), established the framework for the postwar social welfare state and the nationalization of key industries.[1] In Britain, something approaching a national electricity grid had been established following 1926 legislation. After the war, however, the entire electricity supply system was taken under state control.[2]

The template for a larger colonial administrative state in Hong Kong was provided by the experience of the military administration that governed the colony from the return of the British on August 30, 1945, until Governor Mark Young reestablished civilian administration on May 1, 1946. Price controls were put in place for a variety of goods and services, ranging from pork and rice to daily wages for unskilled day laborers. Imports of items, including coal and rice, were controlled by the government. A salaries tax was introduced in 1947. Hong Kong was granted partial fiscal autonomy in 1948, giving it more freedom to run its own affairs. Although Young's proposals for political reform went unrealized, the debate over the Young plan nurtured the development of a more active political society in Hong Kong, including the formation of the Reform Club and the Hong Kong Chinese Reform Association in 1949.[3] The transformation of electricity production and distribution into a political project in turn contributed to the growth and development of civil society. These groups and their leaders played significant roles in the attempt to impose more accountability on the electricity supply companies in the 1958–1964 period.

In keeping with more interventionist policies, a deeper political involvement in providing electricity was seen in many countries. Supplying electricity in Britain, the United States, Hong Kong, and newly independent nations like India became overtly political.[4] Electricity was no longer something to be left to engineers and financiers but became a publicly contested commodity.

Consumer electrification in America expanded in the post-1945 period thanks in large part to regulations that permitted more attractively priced federally insured mortgage financing only for buildings that had higher-capacity electric wiring.[5] Capping a two-decade struggle by Labour, Britain passed legislation in 1947 that nationalized its electricity supply companies on April 1, 1948.[6] This law was to have a significant, albeit indirect, impact on Hong Kong by encouraging the idea that electricity could be subject to political debate and ultimately controlled by the state rather than private companies.

The Japanese seizure of the electricity supply companies and the subsequent British military takeover of CLP and HKE in September 1945 for the first nine and a half months of peace had inserted the government into the electricity supply companies' affairs. State-corporate bargaining over electricity rates and a host of equipment and investment issues had preceded the resumption of civilian control of the electricity supply companies on June 15, 1946. The increased technological sophistication of the electricity supply system coupled with electricity's ever more central role in the colony's economy made government intervention even in peacetime a virtual certainty.

Hong Kong might have reverted to something closer to its prewar small-government model but for one thing: people. The colony's population multiplied sixfold between 1945 and 1960, from about 500,000 to 3.1 million in 1961.[7] Some were Hong Kong people returning after the war, but about half of the new immigrants were born outside of Hong Kong. Demands were made for government action, especially in providing housing. For the first time, ideas about racial equality and citizens' rights, including democratic representation, were widely discussed and debated. Electricity, no longer a luxury for the rich but increasingly integrated into everyday life, came under public scrutiny as never before. There were calls for greater popular participation in the colony's affairs, and citizens demanded more control over electricity—including where, how, and to whom it was provided and at what cost.

The radical transformation of the geopolitical frame in which Hong Kong was situated also worked against a reversion to the prewar model. After the 1949 Communist victory in China, Hong Kong was cut off from its hinterland for the first extended period since the colony was established in 1841. During this period, Hong Kong refashioned itself in the popular imagination as the "West Berlin of the East," a colony that stood on the front lines against a giant Communist neighbor to the north. The existential threat to the colony posed by the PRC reinforced the colonial government's move toward greater state involvement in Hong Kong; it appeared to justify measures that, before 1941, would have been considered extreme. This period should properly be seen as one that laid the foundation for more activist government in Hong Kong rather than as representing some sort of high-water mark of laissez-faire policies.

The New Territories lease was, as of the end of 1947, halfway to its expiration. Observers foresaw that the end of colonial Hong Kong would come with the expiry of the New Territories lease in 1997, if, indeed, the colony managed to forestall a more immediate Communist Chinese takeover. During a lecture tour to the United States in 1954, Governor Alexander Grantham, whose ten-year tenure from 1947 to 1957 made his administration second in length only to that of Murray MacLehose's in the 1970s, was blunt about the impossibility of indefinite British rule:

> In reply to questions as to how long Hong Kong would continue as a British Colony, I would say that, in my opinion, 1997 will be the fateful year, for in that year the lease of the New Territories runs out, and I could not conceive of any Chinese government of whatever complexion renewing the lease. Nor could I imagine the rump of the Colony—the island of Hong Kong and the tip of the Kowloon Peninsula—continuing to exist as a viable entity, with the great bulk of the water supplies coming from the New Territories, and the dividing line between the leased and the ceded parts running right through the new runway at Kai Tak, the airport.[8]

The British government was keenly aware of its tenuous hold on the colony. Hong Kong was classified as a "problem" territory "merely held by the grace of Mao Tse-tung" in a December 1958 Commonwealth Relations Office exchange.[9] Grantham and Kadoorie were among those who saw that an arrangement would have to be made with the Communist government in China well before that date if Hong Kong's prosperity was to be assured. Speaking in 1959, Kadoorie summed up Hong Kong's dilemma: "The essential degree of permanency is lacking in this Colony."[10] Kadoorie was thinking of the stability needed for long-term investments of the sort that CLP made. A lack of permanency was also reflected in an immigrant population that regarded Hong Kong not as home but as a waystation.

The Korean War underscored Hong Kong's precarious geopolitical position. Embargoes imposed by the United States in 1950 and the United Nations in 1951 shut off much of Hong Kong's trade with China, depriving the colony of what had been its primary purpose, that of a port serving China.[11] The Chase National Bank, long associated with the Rockefeller family, was among those that left during the Korean crisis. Kadoorie cited Chase's flight from the colony as evidence of "the uncertainty of conditions in Hong Kong."[12]

Hong Kong was, in Grantham's words, "an innocent sufferer" in the Korean conflict.[13] The territory was on the edge of a fault line, one that separated a colonial, capitalist enclave from an anti-colonial, revolutionary Communist continental power. These geopolitical tensions heightened the colony's sense of worry over how to accommodate and assimilate Chinese refugees in Hong Kong. Endacott, writing in the 1950s, described living in a city caught up in a global struggle. After China sent troops to help North Korea in 1950, the United States imposed a trade embargo in December 1950, and the United Nations followed suit in May 1951. As a result of the embargo on China, the embattled colony in 1951 was termed a "dying city" by a visiting American journalist.[14]

The West Berlin analogy notwithstanding, daily reality was less one of confrontation than of accommodation. Unlike in Berlin, where an Allied airlift was needed to supply the city with food and coal, Guangdong authorities cooperated with their Hong Kong counterparts to provide food and did not impede deliveries by sea or air.[15] Grantham noted that the colony slaughtered more pigs than any city in the world outside of Chicago and that 90 percent of the animals came from China.[16] In the 1960s, Chinese-supplied water eased the colony's ongoing water shortages. Hong Kong in turn was useful to China, providing significant foreign exchange revenue.[17]

A "GREATLY INFLATED POPULATION"

Were Hong Kong's people an asset, a hard-working, entrepreneurial group thriving under the light touch of British colonial administration and its promise of law and security? Or were they a problem, particularly the more than one million refugees who, in the eyes of many of the elite, threatened to overwhelm the colony? Colonial officials, led by Grantham, saw people as a problem to be managed, a mass to be housed, fed, and controlled. The first chapter of the official Hong Kong government annual report in 1956, written by Acting Colonial Secretary Claude Burgess, used Grantham's "problem of people" phrase to sum up the difficulty facing the colony.[18] So important was this issue that the chapter was separately reprinted as a book.[19] This book displays the anxiety of administrators trying to govern a fast-growing Chinese population of almost three million in a colony where non-Asian civilians numbered about 25,000.[20]

For colonial authorities, the refugees represented disease, disorder, and danger. The city's sprawling informal settlements housed more than 600,000 people. These were typically simple one-story wooden houses with corrugated metal roofs and no running water; some had illegal electrical connections. Another 300,000 lived in substandard housing, crammed into makeshift rooftop dwellings and subdivided apartments.[21] Tuberculosis was a particular source of concern, as evidenced by weekly tallies of new cases reported in newspapers. In 1959 about 60,000 people, or 2 percent of the population, had active cases.[22]

Fires fanned public fears about the dangers posed by refugees and their living conditions. A blaze in a Kowloon squatter settlement in January 1950, up to that time the worst in the colony's history, made 20,000 people homeless. A fire on Christmas Eve 1953 in the northern Kowloon Shek Kip Mei squatter settlement burned six villages and left 53,000 people without homes. The government noted that refugees cooked their food with charcoal and wood and some of their huts were lit with kerosene lamps or candles. Chickens, dogs, and pigs shared the huts or lived in the narrow lanes. "In such conditions, every kind of vice flourished," observed the report. "Drugs were manufactured, sold and stored; there were divans, brothels and gambling houses; every form of crime was sheltered by the anonymity of these dark places."[23]

The government had been reluctant to provide housing for fear that this would imply responsibility for schooling and health as well. There had been hope that the private sector could solve these problems, but the scale of the need was far beyond the financial or organizational capacity of housing societies, and financially unattractive to private developers.[24] The Shek Kip Mei fire galvanized the government, which quickly decided to build as much housing as possible on the site of the fire.[25] The official government yearbook called the Shek Kip Mei fire an "act of God" for the forty-five acres of buildable land it opened up in a city where land that could be developed for housing was scarce. The land was cleared with help from the military. Authorities committed $250 million of public funds to build and finance resettlement buildings and take responsibility for feeding the homeless at Army-run soup kitchens until they moved into the new homes.

In July 1954, a fire at the Tai Hang Tung shantytown, not far from Shek Kip Mei, made another 24,000 people homeless, underscoring the need for action. By the following March, less than eight months later, eight seven-story blocks with a total of 4,606 rooms had been built on the Tai Hung Tung site. Each of the 120-square-foot rooms held on average five adults. It was an impressive physical and administrative accomplishment for a government that had long avoided any role in housing and most other social issues. Still, in late 1955, it was estimated that there were about 500,000 squatters in Hong Kong.[26] When *A Problem of People* was published, there were an estimated 334,000 squatters remaining. Rehousing them would require some $200 million in public funds, 200 acres of land, and 120 seven-story apartment blocks.[27]

Riots in 1952 and then in 1956 contributing to calls for more forceful action to solve housing and other social problems. October 1956 saw the deadliest political violence in the colony's history, with some sixty dead, mostly at the hands of police, in riots initiated by triad gangs loyal to the KMT.[28] A detailed official report was prepared. Grantham's five-page cover letter, addressed to Secretary for the Colonies Alan Lennox-Boyd, articulated the colonial authorities' sense that the concentrated urban area contained an ever-present possibility of mob violence.

Grantham worried that because of the "dryness of the timber," what he termed "mob emotions" can be "ignited by any suitable spark." Refugees had "produced conditions of unparalleled overcrowding and the attendant threats to law and order that arises thereon." The "disorders" were "exploited for their own purposes by gangs of criminals, hooligans, and Triad societies." The riots confirmed colonial authorities' concerns about the impact of dense living conditions on an unruly, politically divided, immigrant population. Grantham noted that the density of the living conditions, "which must be about the highest in the world," contributed to the difficulty the police had in containing the violence and observed that nearly half the reported crime in the colony was committed in the northern part of Kowloon, where most of the rioting occurred.

The *Report on the Riots in Kowloon* gave voice to colonial authorities' chronic fear of losing control of the colony's populace. Grantham, in his cover letter to Lennox-Boyd, noted that "closely packed squatter huts dotted all over the urban areas of the Colony" created "a very serious fire and health hazard." By the time Grantham wrote, 200,000 people had been resettled since the Shek Kip Mei fire three years earlier. Worryingly for the authorities, however, 125,000 of them were housed in the sorts of seven-story blocks that were at the center of the disturbances. Simply building new housing did not solve the underlying problems.

Grantham explicitly tied the use of new technology to the state's ability to exert much-needed control. Police communications needed to be improved, he counseled, in the wake of the 1956 riots. Police were not in touch with headquarters because they had no way to communicate. Yet technology brought its own problems, attracting attention and prompting further protests. Grantham noted that police radio cars were a special target of rioters. Although important for quickly transmitting information about demonstrations, the radio cars could not stay at riot scenes, since they would have needed too many police to protect them.[29] Technology in this instance threatened to establish a negative dynamic, with needed devices of control provoking a strong counterreaction.

Grantham contextualized Hong Kong's global position by noting that the world had focused disproportionate attention on the plight of some 100,000 refugees from Hungary while doing little for the five to six times that number who came to Hong Kong in 1949–1950. He noted that most of these immigrants lived "at a bare subsistence level; they own little or no personal property; the great majority have no real stake in the Colony." With jobs hard to come by and "indifferently paid," a "sense of frustration and bitterness" prevailed, leaving Hong Kong at risk of another similar riot.[30] Kadoorie's focus on providing electricity to power factories that would produce both competitively priced goods for export markets and jobs for hundreds of thousands of on-the-edge workers, offered a partial solution.

Historically, Hong Kong had prospered thanks to the free flow of people and goods through the colony. Much of the flow of people was quotidian and prosaic, involving travel between the colony and neighboring Guangdong—routine movements back and forth across a porous border. Hong Kong had long been a node in China's riverine and coastal junk trade.[31] Historian Elizabeth Sinn has shown how Chinese from Guangdong and further afield had long flowed through the colony on their way to California's gold fields, Canada's railroads, and Malaya's tin mines and through the colony again on their way home, whether alive or dead.[32] Other than the period from the Japanese invasion of Guangdong in 1938 until the end of the war in 1945, the border since 1841 had generally been open. That changed in 1951 when strict border controls were put in place to stop the continued flow out of the mainland following the 1949 revolution. Kadoorie was a member of the Legislative Council and was present on June 27, 1951, when the closed border area law was passed into law.[33] Hong Kong was shut off from its hinterland, stanching the flow of people that circulated into and out of the colony.

Shutting off the flow of people and merchandise posed an existential threat to Hong Kong. A closed border eliminated the original reason for the colony's founding as a protected place on the edge of China that allowed relatively unhindered engagement with the world's most populous nation. Now, rather than facilitate the flow of people and goods to and from China, Hong Kong needed to chart a more autonomous path. In retrospect, it was an overstatement to call Hong Kong a "dying city," but it certainly looked like a dead-end one, a collector of people who were trapped in a cul-de-sac.

The idea of Hong Kong as the emporium for products from China had to be abandoned. Hong Kong needed to be reenvisioned, reimagined, rethought. Refugees sneaked into Hong Kong, slipping across the border by land or sea, hidden in sampans or taking the risk of swimming, while train, plane, and ship services that had connected the colony with China had all but halted. Colonial maps displayed blank space north of the border. The territory was linked to the rest of the world only by the tenuous threads of shipping and airplane routes, postal services, and telegraph and telephone networks.

The United States increasingly took over Britain's role as a market for the colony's manufactured goods, as a source of investment, and as a guarantor of security. The importance of the U.S. military, especially its Pacific Fleet, increased after Hong Kong's naval dockyard saw its nine-thousand-strong workforce reduced by one-third in 1948 and the yard closed in 1957.

Despite the challenges of a large refugee population, postwar Hong Kong had better prospects than many other cities in East Asia. In early 1948, the prominent U.S. magazine *Collier's* declared that the good times were back in Hong Kong, in an article headlined "Hong Kong—Boom Town." The subtitle read: "Little damaged by the War, Hong Kong is now the glamor spot of the Orient, replete with business moguls, taxi dance halls, flourishing pickpockets and an opium traffic which is still thriving despite the police." The text of the article continued in a similar vein:

> Hong Kong is the boom town of the Orient today, the only city in the Far East approaching the bright-light bustle of the "good old prewar days," and its British stability has lured thousands of wealthy Chinese families as well as thousands of hungry coolies from what the colonials here for a century have been calling "the troubles in China." That stability has also quieted, at least temporarily, Chinese demands for the return of Hong Kong to China.

As long as the civil war rages and economic chaos impends, most Chinese seem to think the Union Jack looks all right over the City of Refuge. . . .

After the war ended, and Hong Kong was taken back from the Japanese, the British quickly restored law and order, efficient civil administration, a stable currency plus an adequate if partly rationed food supply and relief rations for the jobless.[34]

The article mixes admiration for the way in which a combination of efficient British administration, stable security, and ample supplies of food had produced a Chinese society that existed nowhere else in China with warnings about what lay beneath the city's cosmopolitan, capitalist veneer. Author Weldon James, an experienced Asia correspondent, approvingly noted that items like nylon stockings, California oranges, and canned fruits sold below government price limits, a testament to an ample supply.[35] The postwar context of state involvement in the economy is implicitly underscored by James's lack of further comment on the wide-ranging government price controls imposed even in free-wheeling, almost-anything-goes Hong Kong. The article's lengthy section on the colony's opium dens and its floating, unstable population was a reminder that alongside the colony's "law and order, efficient civil administration, [and] a stable currency," lurked forces of chaos, disorder, and criminality.[36]

Problems in the agriculture of the New Territories and elsewhere notwithstanding, colonial Hong Kong was a shining city on the harbor, a beacon of light on the Chinese coast that now was, with the exceptions of Hong Kong and Portuguese Macao, governed by the PRC. John Keswick, writing in a 1957 book on Hong Kong's economy, marveled at "its gay neon signs, its shops filled with merchandise from all over the world, air-conditioned and bright with fluorescent lighting." He approvingly cited the notion that "in a Far East overshadowed by the curtain between two ways of thought and life, Hong Kong is the only place in the Far East in which the great freedoms can still be understood and practiced, the only bridgehead over which they could be carried into China."[37]

The more than one million new residents gave a sense of new possibility. Although the prevailing narrative was the danger posed by the refugees, there was a more positive discourse as well. With the state providing housing and food to those who had been burned out of their homes, room for more positive social experiments might be found. The two sides were summed up in *A Problem of People*: "The recent riots [in October 1956] alone provide an indication of the possible threats to public order that may be sparked off and spread like a forest fire in such conditions. And it would be criminally foolish to overlook the opportunities which these estates offer to political or even subversive agitators. On the other hand, such compact and uniform communities probably present a unique field for experimental education in the social and civic spheres."[38] The colonial government answered what it saw as a "problem of people" with energetic state action of a sort never seen in the century-old colony, taking on the opportunities provided for social and infrastructural experimentation. By 1960, nine new resettlement estates with accommodation for 270,000 people were built.[39] The water system was expanded in 1957 with the opening of a new reservoir at Tai Lam Chung, the first built after the war and one that had half as much capacity as the colony's other fourteen reservoirs combined.

The list of unprecedented government initiatives was lengthy. Army soup kitchens fed the homeless. A new runway at Kai Tak airport was completed in 1958 and a new terminal building, which contained 115 miles of electric wiring, was finished in 1961. By 1959, government hospitals had vaccination programs against typhoid, cholera, rabies, tetanus, and plague, while BCG vaccinations to protect newborns against tuberculosis were dispensed free by doctors and midwives throughout the colony.[40] Queen Elizabeth Hospital, whose cornerstone was laid in March 1959 by Prince Philip, was the largest hospital in the British Commonwealth.[41] By the time of the 1961 annual report, a government administrator was confident enough to quip that "Hong Kong has not so much a problem of people as a people with problems."[42]

No longer would the state simply impose order, uphold property rights, and administer colonial justice. To maintain its authority and its ability to rule at a time of heightened political and social aspirations among an expanded population, the state would have to exercise more control and take greater responsibility for many aspects of its colonial subjects' lives, from housing to food to schooling. Robert Brown Black succeeded Grantham as governor in 1958. In an address to the Legislative Council on March 6, 1958, Black's first topic was the issue of the colony's population. Referring to "certain tasks of a monumental, as well as of a momentous, nature which relate directly to our greatly inflated population," Black made the case for continued government action:

Sir Alexander [Grantham], last year, commented particularly on this difficult question and described it as a problem of people. Whichever way we turn in reviewing our commitments we come slap up against it. Just as I am certain that the policies were rightly taken in accepting the necessity for planning and expenditure arising out of the presence amongst us of so many people, so I am convinced today that we must go on accepting these commitments. To stop in the middle now would mean that we should go off at half-cock and this could have serious results on our economy and on our social life.[43]

Black also spoke at length about water, tuberculosis, schools, and the opening up of Clearwater Bay peninsula to the public after many years as a restricted military site.[44] There would be no retreat from more activist and interventionist government policies.

NEW OPPORTUNITIES AND NEW DANGERS FOR CHINA LIGHT & POWER

It was against this backdrop that Kadoorie rapidly extended his electricity supply system. Growth rates for electricity use in the 1950s and 1960s were extremely high, often more than 20 percent a year, with consumption doubling in as little as four years. CLP's policy of expanding generating capacity ahead of actual demand made this growth in consumption possible; the ample electricity supplies in turn made possible further increases in economic activity, as well as educational opportunities (schools in crowded Hong Kong often ran double shifts, operating from morning until evening), and leisure activities.

CLP served a more active state in its efforts to bring urban order out of the chaos of a large refugee population. At the same time, electricity provided new, modern, urban opportunities, opening up previously unimagined possibilities for businesses owners, workers, and consumers. On a fundamental level, the physics of electricity, of harnessing the flow of electrons, means that electricity is about control. Electricity needs to be handled in a careful, methodical fashion or it can kill. Be disciplined with electricity or be punished by it. The electricity system established by Kadoorie fit into a pattern of colonial observation and control. Electricity meant shining lights in the darkness and banishing anonymity. It meant CLP engineers and

"A PROBLEM OF PEOPLE"

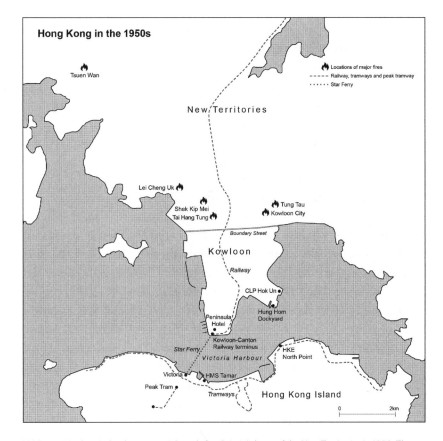

MAP 4.1. Kowloon's development quickened after Britain's lease of the New Territories in 1898. The Kowloon-Canton railroad started service in 1911. The Peninsula Hotel opened to the public in 1928, part of the Kadoories' strategy of building their businesses in Hong Kong to balance those in Shanghai. The takeover of CLP in the 1930s and the decision to expand the Hok Un generating facility were key components in this balancing. Development in Kowloon and the New Territories accelerated after 1945, as the colony's population sextupled over fifteen years when returnees who had left during the war were joined by more than 1.5 million refugees from China. Large squatter settlements, which were susceptible to fire and disease, were replaced with public housing. CLP provided the electricity for factories, apartments, and shops. CLP's electricity powered the colony's stunning postwar economic boom. The expanded population provided workers and consumers who furthered Hong Kong's economic growth.

meter readers going into the Kowloon Walled City, which was off-limits to colonial officials. Similarly, CLP employees went into squatter settlements, where few officials ventured. Electricity was the quintessential expression of a technologically determined new order, one that was efficient, hygienic, and modern.[45]

Conversely, another part of the electricity system involved harnessing and channeling energy, both human and electrical, in a way that allowed for different ways of living and working and in so doing facilitated high and sustained economic growth. Paradoxically, control enhanced the freedom of Hong Kong people to explore new opportunities in work, leisure, and education as owners, producers, and consumers. Electricity literally brought with it enlightenment, incandescent and fluorescent light that allowed students to read at night, as well as mechanical power that drove industrial machines. Electricity multiplied and enhanced human and mechanical powers in an increasingly populous city.

Electricity had been the subject of state control and political contestation in the nine months of military administration in 1945–1946. This was the beginning of what became sustained state involvement in overseeing Hong Kong's electricity supply system. Authorities signaled a new assertiveness in 1948, when the government peremptorily ordered the electricity companies to cut the rates they charged for electricity. A letter from the Colonial Secretariat dated November 6, 1948, advised CLP to cut lighting rates by ten cents a kilowatt hour (KwH) to thirty cents a unit, a 25 percent reduction, and to reduce power rates by two cents a KwH to ten cents a unit, a 17 percent cut. The government also demanded that dividends be restricted and in a November 9 letter discussed the need for legislation to implement the change. On November 19, Colonial Secretary David MacDougall, who had been the military administrator in 1945–1946, called utility heads together in a confidential meeting to discuss limits on dividends and the imposition of legislative controls on the electricity-generating companies. The government paper prepared for the meeting noted that "the legislation contemplated is a trend of the times and Government were not going to push the matter or be in a hurry." The phrase "trend of the times" almost certainly refers to the nationalization of the British electricity supply industry in 1948.

CLP's chairman Albert Raymond expressed "disappointment and surprise" at the government request. Raymond added that the CLP board "cannot find words to express their sheer astonishment" at the suggestion that annual dividends be limited to 12 percent of paid-in capital. Raymond warned that if dividends were restricted on all utility companies, such as the Star Ferry and bus companies, "it would smash the market and bankrupt a large percentage of the investors who have contributed in no small measure toward the rehabilitation of the Colony."

Raymond summed up the difficulties CLP had overcome and stressed its role as a private company that acted in the broader public interest. CLP had paid no dividend for its first eighteen years or during the war years and had given out no bonus or founders' shares. Raymond claimed, seemingly erroneously, that at the start of the war the Royal Engineers blew up the company's high-pressure turbine but that CLP had not received any compensation. He stressed that CLP had cut tariff rates six times since 1946 and that power costs in relation to manufacturing were "well-nigh negligible." Raymond added that the company had an expansion plan that would cost about one million pounds sterling over the next four years before warning: "Government's intervention will undoubtedly strangle, if it does not kill, all enterprise and initiative." Raymond also noted the insecurity of the company's fuel mix and the lack of control over pricing. CLP had switched from coal to oil as a fuel source. If it was forced to use coal again (owing to a lack of oil), that change alone would add five cents per KwH to CLP's generating costs.[46]

Raymond was a member of CLP's board of directors from 1938 to 1955, serving as chairman from 1945 until his death in 1955. Like Kadoorie, he was a Baghdadi Jew with strong ties to the Sassoons, working in Hong Kong since 1934 as the managing director of the E. D. Sassoon Banking Co. Ltd., established by Victor Sassoon in 1928. He was the president of the Hong Kong Jewish Refugee Society in 1938, helping to raise funds through its bulletin and a "buy Jewish" campaign. Shortly after the war he was the colony's exchange controller.[47]

The CLP board in 1948 consisted of five people: Raymond, Lawrence and Horace Kadoorie, M. K. Lo, and Harold Dudley Benham.[48] Three of these were prominent members of the Baghdadi Jewish diaspora, and Benham was a manager working for the Sassoon family. Hong Kong's Jewish community was small. In 1946, Kadoorie put the number at about one hundred people just before the war, of whom about 30 percent were what he called Sephardic, which would have included Baghdadi Jews. He counted just about fifty-five Jews in 1946.[49] The Baghdadi Jewish presence on the CLP board contributed to the anti-Semitism that underlay the attack on Kadoorie and CLP in the late 1950s.

Anti-Semitism in Hong Kong was pervasive. At a dinner Grantham hosted in the 1950s, he hid from the other guests the fact that Elizabeth Taylor's husband Michael Todd was Jewish.[50] During the electricity debate, a Kadoorie defender in the *SCMP* hit back against apparent anti-Semitic whispers. The newspaper's financial correspondent complained about the "misguided public criticism" directed at the company director, "who ensured that the power was there in abundance for every industrial and civic need, even if the water wasn't!" The financial correspondent continued: "The name of the director is used far more often than the name of the company, and by people who should know better. This denigration has gone too far."[51] Four years later, in 1967, Kadoorie was forced off the HSBC board because he was Jewish, as the bank bowed to anti-Semitic pressure in the Middle East.[52]

The government abandoned its 1948 attempt to control electricity prices and dividend payouts. The plan to control electricity supply companies may have been "a trend of the times," but the Hong Kong government in 1948 proved unwilling to embrace this trend in the face of CLP's resistance.[53] The absence of actively engaged manufacturing interests or civil society coupled with the exigencies of postwar reconstruction and the Kadoories' proven expertise in constructing and operating a growing and reliable network meant that colonial officials were unable to effect any systemic change in the prevailing electricity supply arrangements. There may also have been the problem of administrative capacity. The increased flow of refugees into the colony as the civil war in China drew to a close preoccupied officials and may have contributed to their abandonment of this politically controversial initiative.

After the 1948 attempt at control was rebuffed, the electricity supply companies reverted to their prewar pattern of running their businesses autonomously, with little official interference. Nonetheless, there were unofficial conversations between Kadoorie and the government regarding electricity supply over the next decade. During the ESCC hearings in 1959, Kadoorie repeatedly cited a conversation he had had with Governor Grantham and Financial Secretary Arthur Clarke in 1954. Kadoorie testified, and there is no record of any denial (or confirmation), that the three men explicitly discussed electricity pricing, cross-subsidies by consumers for industry, and the need to increase electricity-generating capacity in order that factory employment could remain high. Kadoorie also appears to have written the government a letter on this subject in 1955. (The letter has not been located.)

Although anti-Semitism added fuel to the campaign against CLP, the company's growing profitability and need for additional capital would likely have made it a political target in any event. Within two years of the war's end, CLP was earning record profits. War damage was less than initially feared, and demand for electricity was high. Profits had already been increasing in the two years before the war because of the capacity of the new generating plant at Hok Un to produce larger quantities of electric current. For the year ending September 30, 1941, the first full year that the expanded Hok Un station was in operation, profits totaled a record $1,241,255. In the 1947 fiscal year, earnings were $4,122,978, more than triple the prewar high. Three-quarters of those profits were paid out in the form of dividends. Two years later, in 1950, profits doubled again to $9.5 million. By 1960, profits (or "income from the working account," since CLP presented only fragmentary financial information) had quintupled to $48.5 million, thirty-nine times the prewar record established in 1941 after the Hok Un expansion.[54]

CLP continued to expand throughout the 1950s. In this it followed Kadoorie's policy, laid out in his 1938 wedding day manifesto, to buy good equipment and build capacity in advance of actual demand for electricity. "Every effort will be made to ensure that supply is readily available in any part of the territory served by our undertaking," the company promised. CLP struggled to keep up with the growth in electricity demand in a world in which the production of new electricity-generating equipment was constrained.

Kadoorie noted that there were numerous difficulties as a result of the increasing appetite for electricity, "but despite these I can with satisfaction report that it has not been necessary either to restrict or deny supplies to any consumer."[55] CLP, in its annual reports, and Kadoorie, in his public statements, repeatedly affirmed that the company was ready to supply power and light to anyone. To take one of many examples, in the 1954 annual report it noted that it had run a cable to a Ma On Shan mining operation. CLP's approach was markedly different from that of HKE, which was often unable to supply needed light and power.

CLP's expansion had long been driven by industrial users, not consumer or commercial ones, reflecting Kadoorie's interest in industrial development.

In 1935, two decades before Hong Kong was known as a manufacturing center, Lawrence Kadoorie wrote shareholders that throughout Kowloon and the New Territories there were "no less than 86 different kinds of industries served by this company," ranging from shipbuilding to toothpick manufacturing, from cement works to cigar factories.[56] CLP's 1955 annual report noted 604 new factories.[57] Cotton-spinning and related industries, such as weaving and dyeing, were the most important sectors.

The annual report noted that the company was supplying electricity to industrial establishments ranging from bean-curd makers to stone quarries to watch-case manufacturers. By 1955, over 230,000 people, or about one-third of the working population, was employed in factories of fifteen or more people. At least 100,000 more people worked in smaller, unregistered factories.[58] Some of the larger factories were modern establishments.

A visitor to one of the colony's cotton mills in the 1950s was, according to John Keswick, "astounded to find that it was equipped with the most modern machinery and operated under the most hygienic conditions."[59] While Keswick's account of his unnamed informant may be accurate, it can also be read as an attempt to forestall growing criticism in Britain about conditions in Hong Kong's factories and the related threat to limit imports from the colony. Keswick was a director of Jardine, Matheson, a group that helped provoke the Opium War and with which his family had been associated since the early nineteenth century.[60]

In the immediate aftermath of the war, Kadoorie prepared the way for Shanghai-listed companies to move their domiciles from Shanghai to Hong Kong.[61] As the Shanghai textile manufacturers and other mainland Chinese companies began to think about moving, Kadoorie made it known that CLP was ready to provide all the electricity they needed. As one of the leading business families in prerevolutionary Shanghai, the Kadoories were particularly well-placed to tap into the Shanghai textile manufacturers' and other industrialists' need for new factory sites. Kadoorie's success in wooing industrialist Y. C. Wang, a friend of his from Shanghai, was typical in its approach. Michael Kadoorie says that shortly after the war Wang asked his father: "'If we come to Hong Kong can you provide us with electricity?' My father promised electricity but it was a hollow promise, of course, the facilities having been destroyed during the war."[62]

Kadoorie's promise certainly was audacious. Hok Un had been repaired by this time and the problem was no longer war damage but trying to keep up with increased demand. In the event, Wang took up Kadoorie's invitation. The Nanyang Cotton Mill, which Wang founded in 1947, was both one of the first and one of the largest cotton mills to be established in the colony. It started operations in 1948 with 15,000 spindles and 200 looms. At its peak, Nanyang cotton had 55,948 spindles and 547 looms.

Lawrence Kadoorie helped Nanyang get started by selling the company land for its new factory in Ma Tau Kok. The property had been owned by another Kadoorie company, Hong Kong Rope Manufacturing, founded in 1883 and originally part of the Shewan Tomes stable of companies. The site was requisitioned by the Royal Air Force after World War II. Kadoorie negotiated the return of the land so that it could be sold to Nanyang in 1947.[63]

The strength of the Kadoories' connections with the elite Shanghai business community was demonstrated by the formation of the partnership with Wang. Kadoorie was chairman of Nanyang Cotton Mill from its founding in 1946 until his death in 1993. Wang served on the CLP board of directors from 1960 to 1992. Wang was also a founding director of another Kadoorie interest, Tai Ping Carpets International.[64] Wang's wife was the niece of a prominent Shanghainese industrialist, Rong Yiren, who later became vice president of the PRC and played an important role in China's economic opening of the 1980s and 1990s.[65]

"It was not a foregone conclusion if they would come to Hong Kong," says Michael Kadoorie. "[Wang] and others in the textile industry were considering places like Brazil and Mauritius so electricity really made the difference in terms of them coming [to Hong Kong]."[66] Helping tilt the decision in favor of Hong Kong was Kadoorie's ability to supply land in 1947. Later, the relatively large supply of land available for factory construction in the New Territories, some of which the government supplied in the form of industrial estates, drew other industrialists.

Up to two-thirds of Shanghai's privately owned cotton spinning factories set up operations in Hong Kong during this period. Many of them used modern equipment that was en route to Shanghai but was rerouted to Hong Kong when the decision was made to leave China.[67] Historian Steve Tsang contends that post-1949 Shanghai entrepreneurs gave the colony a ten- to fifteen-year head start over the rest of Asia.[68] Although the scale of this industrial development was unprecedented, it should be emphasized that pre-1941 Hong Kong had a significant manufacturing industry. Tak-wing Ngo notes that one survey found two thousand Chinese-run factories in Hong Kong by the late 1930s, employing more than 100,000 people.[69]

Even after Shanghainese entrepreneurs moved their textile operations to Hong Kong, the government was reluctant to abandon the idea that manufacturing was anything other than a peripheral activity. Grantham in 1949 told the Legislative Council that "Trade is the life blood of this Colony.... I am proud of being Governor of a Colony of shopkeepers." While noting that industry was developing in Hong Kong, he said that its future "remains obscure."[70] In his memoirs, Grantham claimed that the government misunderstood what was occurring, as it thought that refugees would return to China after the political situation stabilized.[71] "This interpretation followed from colonial policy, which dictated that Hong Kong should be an entrepôt, even though Chinese industry was the largest employer in the colony."[72]

TECHNOLOGY AND CHINA LIGHT & POWER

CLP was both one of the territory's most important introducers of technology and a catalyst of technological change in the colony. It bought its generating and transmission equipment from British suppliers and, in the words of Lawrence Kadoorie, proudly "bought of the best." This pattern of heavy investment in high-quality capital stock continued throughout the more than hundredfold expansion in generating capacity that took place from 1945 to 1982.

CLP introduced clusters of new technologies into the colony. Its electricity generation and transmission equipment was the most visible and the costliest. Another important set of technologies related to metering electricity, billing customers, and collecting their payments. A third set of technologies centered on communications, particularly between CLP's headquarters and its Hok Un generating station, and between these fixed locations and its workers in the field.

The technology CLP used was a source of pride to Kadoorie and others at the company. Its 1952 annual report showed a picture of the newly installed No. 10 boiler at the Hok Un generating plant as well as the accounts department's "mechanical billing section," featuring "the modern Billing and Cash posting machines used in the production of Consumers' bills and accounting records." Manufactured by the U.S. National Cash Register company (NCR), the department included three billing machines as well as two typewriters; in the photographs, all were operated by Chinese men.[73] The following year's annual report noted that, while the new mechanical billing system was functioning well despite the addition of many new customers, the company had ordered a new "Public Utility Billing Machine" to meet the needs of continued expansion.[74] A decade later, CLP purchased the colony's first computer, also from NCR.[75]

CLP had modern communications equipment in order to stay in touch with crews working remotely. The company's transmission and distribution network covered more than 1,000 square kilometers of the colony's approximately 1,100 square kilometers.[76] VHF radio was installed in 1950 to communicate with crews in the New Territories and, in 1957, Lantau.[77] The company repeatedly stressed the usefulness of its new technologies to shareholders in its annual reports. "The Radio Telephone system again proved its worth throughout the year under review, and particularly minimized loss of time when attending to breakdown work consequent upon Typhoon 'Susan,'" read a characteristic assessment.[78] The 1954 annual report has a photograph illustrating the use of radiophones in everyday working situations, noting that they could help dealing with emergencies and that they were in the main station at Hok Un as well as in trucks.[79] The following year's report predicted that the VHF system would be particularly useful in Lantau, "where no alternative system of communication is available."[80]

The electricity supply companies were acutely aware of the difficulty of producing reliable and expanded supplies of electric current in a location halfway around the world from the suppliers of what was still an unstable technology. The CLP 1953 annual report referred to "manufacturing problems" with the installation of the No. 10 Turbine at Hok Un and noted the "unsatisfactory progress in regard to this unit.... The Erection Engineer for this Turbine will shortly be arriving in the Colony from the United Kingdom."[81] A more extended lament is found in the China Light 1957 annual report:

> The tremendous expansion of the Colony, both industrial and residential, and the fact that we are dependent for deliveries of machinery and replacements on the United Kingdom has made the problem of maintaining a continuous and unlimited supply difficult to resolve. Orders for plant of the size

necessary for our Company's development take two to three years to fulfil and, in addition, several months are required for erection and commissioning on arrival in the Colony. Our programme was based on promised dates for completion but unforeseen delays occurred. Although in the postwar period in many parts of the world, including Hong Kong, public utilities have been unable to cope with the demand, your Company rightly prides itself that never yet has it refused supply to industry. The above-mentioned delays explain why plant normally held in reserve has had to be used to the full with consequent interruptions due to lack of standby capacity.[82]

The allusion to "consequent interruptions" constituted a rare admission by CLP that it was sometimes unable to provide uninterrupted supplies of electricity.

CLP tried to overcome the tyranny of distance and its reliance on expertise from Britain by ensuring that it had well-trained technical staff on site. It hired engineers from around the world, mostly those who had been working in other parts of the empire or in Commonwealth countries. Beginning in the 1930s with the Hok Un extension, Kadoorie developed close personal and institutional relationships with the manufacturers of the company's equipment, writing and personally meeting both company heads and their local representatives. Kadoorie often nurtured these relationships with extensive correspondence, ranging from business letters to holiday cards. Kadoorie took an intense personal interest in technical and engineering issues, and the first extended letter he wrote after the end of World War II contained in the HKHP archive was to an equipment supplier.

Kadoorie companies took responsibility for an increasing amount of the construction work, starting with the Hok Un expansion in the 1930s, using Kadoorie-controlled construction firm Hong Kong Engineering Co.[83] The 1960 annual report noted that the company "adopted a most unusual course" by building the Hok Un B station itself, acting as the general contractor and employing direct labour. "We may well be proud of the result, particularly as the overall cost of the new station compares very favourably with that of similar stations in any part of the world."[84]

The electricity that CLP and HKE supplied opened up new possibilities. As noted, electricity made possible the fan and then the air conditioner, as well as domestic appliances such as the refrigerator and the rice cooker. Hong Kong in 1957 became the first British colony to introduce television.[85]

"A PROBLEM OF PEOPLE"

E. Taus, the author of a 1957 essay on electrical appliances, counted thirty-two types of household appliances, mostly powered by electricity, that were sold in Hong Kong. These included massage apparatuses, violet-ray units, liquidizers, meat mincers, waffle irons, clocks, razors, and door chimes. "Each and every one of these is undoubtedly a boon to better, safer and easier living," Taus wrote. One drawback was that CLP supplied 200-volt current, but appliances generally were built for 220 volts. Taus noted: "the voltage here frequently drops to 180 volt [sic] or lower and this causes trouble with automatic toasters, irons, etc."[86]

The air conditioner transformed life in the hot, humid, subtropical colony. "Air conditioning is science's greatest contribution to human comfort in the 20th century," wrote Wilfred Wong in the same series of 1957 essays on Hong Kong's economy. Wong listed the colony's one dozen centrally air-conditioned buildings. These included the Government Secretariat, Kadoorie's Peninsula Hotel, the HSBC headquarters, the police headquarters, and the premises of Butterfield and Swire. Wong noted that before World War II there had only been one building, one bank, and two theaters that were air-conditioned. By the mid-1950s the figures had risen to ten buildings, sixteen banks, and thirty theaters.[87] Another commentator in the same collection, writing on films and cinemas, noted that "in the last few years modern air-conditioning has been installed in all of the better theaters in the Colony."[88] Although Kadoorie focused on industry, he knew that air conditioning could be an important source of new demand.[89]

Kadoorie seized on less obvious technological combinations, showing how CLP's electricity system existed as part of an increasingly complex urban network. A notable example was CLP's response to the growing quantity of cash payments generated by its expanding system. Handling cash and transporting it to the safety of a bank was risky. It was also time-consuming and physically challenging. Customers complained that opening hours for payments offices were too restricted. The need to handle cash more safely and reliably led Kadoorie to encourage HSBC, for which Kadoorie became a director in 1957, to set up branch offices in Kowloon and the New Territories at a time when the bank had few outlets other than its One Queens Road headquarters on Hong Kong Island. One of the bank's first Kowloon branches was opened in 1957, occupying the ground and mezzanine floors of an annex to the Kadoorie-owned Peninsula Hotel, the Peninsula Court. The twelve-story building, completed that same year, was "air-conditioned throughout by a reversible chilled and warm water system, the first of its kind in the colony."[90]

Kadoorie leased space that CLP had bought in anticipation of future needs to the bank for a nominal rent of $1 a year in return for the bank collecting payments by CLP's customers. That solved a problem for CLP while allowing the bank to start its branching with virtually no capital cost. The bank then set up a security company, Guardforce, to physically move the cash between its branches and its headquarters. In turn, CLP later became Guardforce's largest client.[91] Hong Kong's urban growth was facilitated by this sort of innovative cooperation arrangement, one that was only possible for the colony's largest and most sophisticated business organizations.

Televisions, air conditioners, elevators, and escalators helped create a recognizably modern city, though much was made of the continuing contrast between new technologies and old ways of doing things. CLP's Hok Un plant was one of the most capital- and technology-intensive in Hong Kong. Pictures of its construction in the 1930s in a company-sponsored history of CLP show barefoot construction workers.[92] The juxtaposition of a more traditional Hong Kong and the modernity of the CLP system was highlighted by a photo of a woman poling a sampan that carried as its freight a bulky modern billing machine, en route to the Tai O cash collection office. Modernity was framed by tradition in the CLP imagery, as it was in many other images of the colony.[93]

NUCLEAR POWER

Lawrence Kadoorie's longstanding interest in nuclear power was first mentioned in CLP's 1956 annual report, perhaps sparked by the opening of the Calder Hall nuclear reactor by Queen Elizabeth in October 1956. Calder Hall was the world's first commercial nuclear reactor. Kadoorie's letter to shareholders that year noted that nuclear power could "revolutionize" the electricity-generating industry.[94] He warned that the new power source could make the company's existing generating stations obsolete.

The widespread diffusion of commercial nuclear power was a remote possibility during the 1950s, yet Kadoorie found a way to turn the threat that nuclear power would render his coal-fired power plants obsolete into a profitable opportunity. In the 1956 report, he stated that the company was taking an additional depreciation allowance, in addition to normal 7.5 percent depreciation, to guard against the possibility that existing power stations would become prematurely obsolete.[95] Kadoorie does not detail the size of the depreciation, a non-cash charge against reported profits. Although the depreciation charge had no effect on the company's cash flow, the effect of Kadoorie's action was to reduce the growth in reported profits at a time of mounting public unhappiness with the company. In fact, the 7.5 percent depreciation rate was already quite aggressive, implying that the new generating equipment would have a useful life of just over thirteen years, an extremely short time for equipment in an electricity-generating facility.

The following year, 1957, Kadoorie sounded an even more ominous warning about the possibility that coal-fired power plants would be rendered uneconomical by nuclear power. The company took the extraordinary step of writing off one-third of the new Hok Un B generating unit with a special depreciation charge.[96] This action implied that the company expected that the new facility would have a useful life of only three years, an improbably short period of time even if nuclear power was a viable competing technology. The $3.8 million charge cut reported profits by 27 percent, to $10.2 million. The lower profits meant lower taxes and perhaps, as with the previous year, was done in part to dampen any public disapproval of the company. This disquiet with CLP was evident in newspaper articles and editorials as well as correspondence with the government. Depreciation is generally subject to agreement from external auditors and government authorities because of its effect in reducing reported profitability and thus taxes, but I have found no evidence of any internal, let alone public, debate about this unusual charge.

Whatever Kadoorie's motives in taking the accelerated depreciation charge, his interest in nuclear-generated electricity was real. In the years to come, he would push for the establishment of a nuclear-powered electricity generation facility in Hong Kong. He investigated the possibility of a nuclear-powered plant that would both desalinate water and produce electricity. Kadoorie was a significant promoter of nuclear power throughout the latter half of the 1950s. By 1958, however, he had become more concerned about the capital costs of building a nuclear generating station and told shareholders that the next plant would be a conventional thermal generating plant:

Shareholders will no doubt be interested to know that experts in nuclear generation have already visited this Colony and that we are co-operating with others in exploring the possibilities of a nuclear generating station for the Colony. Capital cost per kilowatt of nuclear generating capacity particularly for early stations will be very high. Fuel costs, however, after allowing for the initial charge should be much lower than for conventional stations. In saying this, however, I must stress that generation of electricity in Hong Kong by this means is still some distance off and should not be taken as being likely to affect us in the near future. Any immediate expansion which might become necessary will be by means of a conventional plant.[97]

Kadoorie's dramatic reversal of his assessment in the previous year's annual report, when he was alarmed enough about the prospect of the rapid introduction of nuclear power to take a 33.3 percent depreciation charge, begs the question of what happened and whether or not Kadoorie truly believed in 1957 that existing oil- or coal-fired plants soon would be obsolete. In October 1957, a three-day fire at Britain's Windscale nuclear facility, adjacent to Calder Hall, released radiation into the atmosphere and led to the mandatory dumping of milk produced in the surrounding countryside into the Irish Sea because of fears that it would be contaminated with radiation. That fire occurred ten days after the end of CLP's fiscal year on September 30, 1957.

Kadoorie would almost certainly have known about the fire when he prepared the 1957 chairman's letter extolling the promise of nuclear generated electricity. He did note in his 1958 letter that nuclear generation experts had already visited Hong Kong. Perhaps they educated him in more detail about the economics of nuclear-powered electricity generation.

Nuclear power continued to attract interest, as displayed in a special section of the August 27, 1959, edition of the *Far Eastern Economic Review*. Given Kadoorie's close ties with the magazine, the article "Hongkong—Water and Power from the Atom by 1965?" on the possibility of a nuclear-powered electricity generation and water distillation plant in the colony, was almost certainly written with Kadoorie's knowledge.[98]

In 1946, Kadoorie was one of the four original investors in the *Far Eastern Economic Review*, along with Arthur Morse at HSBC, Jardine, Matheson taipan David Landale, and Karel Weiss, a Jewish refugee from Shanghai who later printed the magazine. Editor Eric Halpern, a Viennese Jew who had come to China in 1932, had previously edited a Shanghai-based business magazine backed by the Sassoon family and other businessmen.[99]

Archival evidence points to Kadoorie's ongoing association with the magazine's editors and journalists, and his ESCC testimony echoed the article. Testifying two months later to the Commission, Kadoorie stated: "It is understood that Government is now investigating the possibility of converting salt water into fresh by means of atomic water, and using process steam to produce electricity as a by-product."[100] So the article almost certainly reflects Lawrence Kadoorie's thinking.

The article contended that a dual-use nuclear-powered water distillation and electricity production plant could solve Hong Kong's twin problems of water and electricity scarcity. An *SCMP* article about the story said that the choice was to spend $800 million to $1 billion for nuclear or continue with conventional fossil fuel–generated electricity at steadily increasing cost:

> The belief was expressed that a nuclear power station in Hong Kong would raise political, siting and health problems. However, it was added, experience had shown that the danger of radioactivity had been considerably reduced, even to the extent of siting a station on Hong Kong Island which was thought to be the most acceptable position from the political angle. There would also have to be co-ordination between the two power companies and the Government on running such a project, and the inquiry into the power companies might have a greater significance than many people had thought.[101]

Note the contention that the power companies' inquiry "might have a greater significance than many people had thought." If this reflects Kadoorie's views, it is one of the earliest indications that Kadoorie was reconsidering his opposition to government involvement in large-scale infrastructure projects, notably electricity generation. This mention of "coordination" with the government dovetails with Kadoorie's marginal comment on CLP critic Charles Barber's later ESCC testimony. In that marginal note, Kadoorie opens the way for government/private cooperation on an electricity-generating unit. He recognized, seemingly for the first time in writing, that government could be useful as a provider of capital, and as a partner to share the risk, for capital- and technology-intensive projects. It appears that Kadoorie's analysis of the capital costs involved in commercializing an unstable technology convinced him that a government partnership might be the best form of risk-sharing.

By 1960, Kadoorie appeared less optimistic about a rapid introduction of nuclear power. In CLP's 1960 annual report, the company noted that: "It is authoritatively stated that the cost of electricity from nuclear power stations of the Calder Hall type may never be competitive with that of conventional thermal stations." Instead, Kadoorie turned his hope to other sorts of nuclear power, citing gas-cooled reactors, fast-breeder reactors, and magneto-hydrodynamic methods.[102]

CLP's subsequent agreement with Esso in 1964 explicitly gave the Hong Kong electricity supply company the right to build a nuclear power plant outside of the joint venture.[103] Later in the 1960s, Kadoorie commissioned concept designs for what a proposed Hong Kong nuclear power plant might look like. In May 1974, the Hong Kong government appointed him as a member of a steering committee on nuclear power. In July 1974, he visited nuclear power plants in Britain and France.

The Hong Kong government report, released in January 1977, recommended against nuclear power on the grounds that its cost was likely to be only marginally lower than that of a conventional plant. After the Hong Kong government rejected the nuclear option, Kadoorie used the expertise that he and CLP had amassed to commission and complete a detailed multivolume study on nuclear-generated electricity that he presented to the PRC, UK, and Hong Kong governments in early 1981.[104] This became the groundwork for the Daya Bay nuclear complex, China's first civilian nuclear reactor, which went into commercial operation six months after Kadoorie died.[105]

FIGURE 1. Lawrence Kadoorie (left) and his brother Horace (right), in an undated photo. Sons of Baghdad-born Eleazer ("Elly") Kadoorie and London-born Laura (née Mocatta), Lawrence concentrated on running the family's diverse businesses, while Horace was best-known for his work with the family's philanthropic activities, such as the Kadoorie Agricultural Aid Association and aid to Jewish refugees. (Source: Hong Kong Heritage Project.)

FIGURE 2. Governor Geoffrey Northcote and Lawrence Kadoorie spoke at the ceremony to inaugurate the new electricity-generating station at Hok Un in February 1940, the first public commemoration of an electricity plant in Hong Kong's history. Republic of China flags flank the British flag. (Source: Hong Kong Heritage Project.)

FIGURE 3. The CLP headquarters building on Argyle Street, completed in 1940, is an unusual example of Bauhaus-style architecture in Hong Kong. The move from the St. George's Building in Central came in tandem with the expansion of CLP's generating capacity at Hok Un and reflected a newfound focus on Kowloon. (Source: CLP, Annual Report, 1951.)

FIGURE 4. A CLP facility at Boundary Street and Waterloo Road celebrates the coronation of Queen Elizabeth II in 1953. (Source: Hong Kong Heritage Project.)

FIGURE 5. CLP extolled its street-lighting program, which in the 1950s involved the installation of hundreds of new streetlights a year in cooperation with the colonial government in order to promote public order and safety. In this Waterloo Road picture, street lighting is an integral part of a modern cityscape characterized by private automobiles, efficient roads (with parking areas defined by a traffic separator), wide sidewalks, newly planted shade trees, and new buildings. (Source: CLP, Annual Report, 1955.)

FIGURE 6. Beginning in the late 1940s, observers noted that Hong Kong had become among the first cities in East Asia to regain its prewar buzz. Neon signs, lit up with electricity supplied by CLP, exemplified modern Hong Kong. Illuminated Hong Kong stood in stark contrast to the darkened People's Republic a short distance away. Neon photographed well and photos like these became a trope, especially as color photography became more common in the 1960s. (Source: CLP, Annual Report, 1958.)

FIGURE 7. Workers celebrate the Chinese New Year holiday at the Hok Un expansion worksite. In an unusual syncretic adaptation of the apparatus of modern technology, the large cable spools function much like a temple wall, with the central one hosting paintings of Chinese deities. Three central protagonists dressed in traditional clothing perform rites; three others, including a musician, are dressed in contemporary Western-influenced clothing. (Source: CLP, Annual Report, 1958.)

FIGURE 8. CLP installed the colony's first computer, bolstering its position as one of the territory's most technologically sophisticated enterprises. (Source: CLP, Annual Report, 1959.)

FIGURE 9. The order and technological sophistication of the CLP system contrasts with photos showing the destructive force of Hong Kong's subtropical climate, the danger of theft, and the attendant difficulty of maintaining an electricity supply network. This photo was published at a time when CLP's independent existence was threatened by nationalization and implicitly underscored the company's success in building and operating a reliable electricity supply network. (Source: CLP, Annual Report, 1959.)

FIGURE 10. A traditional poled sampan is used to transport a cash-receipting machine to CLP's office in Tai O, on Lantau Island. This marriage of newer and older, of Western and Hong Kong, technologies is an example of hybrid or creole technology. That the Tai O payment office was only open one day every two months for payments sparked complaints from residents during the Electricity Supply Companies Commission hearings. (Source: CLP, Annual Report, 1960.)

FIGURE 11. British Prime Minister Margaret Thatcher in September 1982 in Hong Kong, presiding over the official opening of the Castle Peak electricity-generating station, one of the world's largest coal-powered facilities and the source of the largest-ever order for British manufacturers. Lawrence Kadoorie is seated in the center. (Source: Hong Kong Heritage Project.)

Chapter Five

ELECTRICITY AS A POLITICAL PROJECT
1959–1964

In January 1960, following hearings the previous autumn, the Hong Kong government announced that the Electricity Supply Companies Commission (ESCC) had recommended that the colony's two electricity companies be merged and nationalized. Citing what it said was an irreconcilable conflict between the interests of consumers and those of the electricity companies, the commission recommended that the government purchase the two companies and run them for the good of the whole community. To forestall asset-stripping before nationalization, the government at the same time froze any increase in dividend payments by the two electricity companies. This preemptory exercise of state power to control private companies was unprecedented in Hong Kong history.

The contentious debate over how to control Hong Kong's two electricity companies in the late 1950s and early 1960s reveals postwar thinking about state power, the relationship between the public and private sectors, and issues of government and governance.[1] The colonial state was reconfigured in the post-1945 period to include more responsibility for the economy and for society, notably for the health and housing needs of the more than one million refugees. Yet many people in Hong Kong, prominent among them Lawrence Kadoorie, argued for renewed adherence to a prewar principle of minimal government interference. Beginning in 1961, the economy was overseen by Financial Secretary John Cowperthwaite, who to the outside world would come to exemplify Hong Kong's laissez-faire exceptionalism. Ironically, these two defenders of noninterventionist policies were the two principal protagonists in the 1959–1964 period in negotiating a new order based on increased state control and on a mutually reinforcing form of business-state accommodation.

The limits to Hong Kong's stated policy of nonintervention are apparent in the case of electricity. Electricity as a technology and as a business exemplified the sort of increasingly complicated and large-scale system that in the postwar period became central to the functioning not just of the economy but of the entire social and political order. Electricity had become essential, yet its price and availability relied on two private companies whose pricing and supply often appeared erratic and arbitrary—especially CLP. The government decided that it could no longer take a laissez-faire approach toward electricity supply, for the colony could no longer function smoothly without ample and reliable current. The government's need to control the electricity supply companies coincided with the increasingly vocal aspirations of business and individual consumers, for whom the issue of electricity became a significant political rallying point. The debate over control of electricity transformed the two electricity supply companies, which, for their part, turned the controls that were imposed to their advantage and as a result became significantly larger and more international. The Hong Kong government became more intimately involved with fundamental aspects of electricity supply and other parts of the colony's physical and economic infrastructure.

The 1959 ESCC hearings elevated electricity to the status of what I term a political project, a shared endeavor that reflected and contributed to the growth of a more active civil society in the colony. Debates over availability, pricing, and establishment of the generating and distribution systems made electricity a controversial public issue for the first time in the colony's history. Electricity as a political project involved conscious, widespread, public attempts by diverse players to shape the electricity system. Public hearings and extended debate in the colony's newspapers created an unprecedented opportunity for sharply differing views to develop and to be heard. The issues involving electricity helped to delineate competing ideological positions and the different economic and political rationales that underpinned them. A close reading of this history shows the mutually reinforcing development of an expanding electricity system, a growing colonial administration, bolder nonofficial voices in the media and elsewhere, and the changes taking place in the colony itself.

The government was under pressure to act on electricity, as on so much else, by various business groups as well as by reformers. The departure at the end of 1957 of the less interventionist Grantham (who on a visit to the Hok Un generating station five years earlier had written "Hurrah for private enterprise!" in the visitors' book), opened up the way for more direct government control of the electricity supply companies. Robert Black, who had been colonial secretary (1952–1954) and governor of Singapore (1955–1957), succeeded the conservative Grantham in 1958; likely emboldened by the nationalization of Britain's electricity sector a decade earlier, it was Black's administration that in April 1959 announced the establishment of the ESCC.

The debate in the late 1950s and early 1960s over who would make key decisions about the colony's electricity supply networks, who would pay for their expansion, and who would profit from them, brought together disparate groups in favor of greater control over the electricity supply companies. It also paralleled and informed debates taking place about the colonial state's responsibility for housing, education, child welfare, labor rights, and health care. Those in favor of political control over the colony's electricity supply included cotton spinners, movie studios, some business organizations, political reformers, proponents of municipal electricity, and unaffiliated critics who focused on electricity supply and distribution as exemplifying profound defects in the colonial structure of political power and economic control. This loose coalition was arrayed against the colony's two electricity supply and distribution companies, particularly the larger CLP, as well as their allies in the banking, legal, and accounting spheres. The government played an active role in making this debate possible, first by establishing the ESCC, and later, under Financial Secretary Arthur Clarke and his successor John Cowperthwaite, in mediating and eventually resolving the different views by establishing new and far-reaching controls.

The most public manifestation of the emergence of electricity's status as a central political issue was the ESCC hearings, three years after the October 1956 "Double-Ten" riots in Kowloon and six years after the Shek Kip Mei fire. The government faced pressures to be more proactive in meeting the needs of the colony's people. The debate over how to deal with electricity's status as a natural monopoly took place in Hong Kong during the end of a period when similarly heated debates were taking place in Britain, continental Europe, and the United States.[2]

Many newly independent states nationalized electricity and other companies in the post-1945 period.[3] Resource-rich countries attempted to exert more control over their assets in this period; the formation of the Organization of Petroleum Exporting Countries (OPEC) in 1960, by Iran, Iraq, Kuwait, Saudi Arabia, and Venezuela, exemplifies this tendency. China had nationalized all industry after 1949. These radical shifts formed part of a broader questioning of other types of private control.

An extract of a letter from a Quebec-based acquaintance of HKE's general manager, William Stoker, analyzed the rise of René Lévesque of the Parti Québécois and his program to nationalize and expand Hydro-Quebec. Stoker's unnamed correspondent noted the "strong wave of nationalism" surging through Quebec. The writer put this down to "modern communication, more general education, and more freedom from Church control. *The nationalization of power utilities is a sort of safety valve.*"[4] So, too, in Hong Kong, was the debate over electricity supply arrangements a kind of "safety valve," relieving political pressure that could have been directed at the government itself.

The framing of electricity supply as a political project demanded significant financial and administrative resources on the part of the government, CLP, and HKE. This process spanned at least five and a half years, from the announcement of the ESCC in April 1959 (records detailing the administrative process that took place before the announcement have not been found) until the simultaneous announcement of the Scheme of Control Agreement (SoCA) and the colony's largest-ever foreign investment by oil giant Esso in November 1964. For CLP, the struggle to keep both ownership and management control in Kadoorie hands required time-consuming efforts on the part of, among others, Lawrence Kadoorie, members of CLP's board of directors and senior management, outside counsel, an economist from the London School of Economics, and senior executives at Esso.

Even most critics agreed that the companies' performance compared favorably with their counterparts in other countries and with other Hong Kong utilities, such as the telephone company and waterworks. The colony's two electricity companies were supplying electricity reliably at rates that were being reduced even as the companies were extending electricity distribution to a rapidly increasing number of businesses and households. It is precisely because this significant moment in mid–twentieth century Hong Kong history is difficult to understand that we should inspect it more closely. A thorough analysis of this fraught moment can, in fact, tell us much about the constellation of forces that made up the Hong Kong of the late 1950s and early 1960s.

THE POLITICAL CONTROL OF ELECTRICITY

Electricity supply companies from the early part of the twentieth century were widely recognized in Europe and the United States as natural monopolies.[5] Natural monopolies occur in businesses where high initial costs—such as the construction of an electricity distribution network—make it almost impossible for a new company to challenge an incumbent. Existing companies thus have an ability to set prices without having to worry about competition. Accordingly, these types of businesses were generally either owned by governments or subject to more stringent and far-reaching state controls than most other businesses. Water and telecommunications were also considered natural monopolies and subjected to similar administrative and regulatory restrictions. All these businesses required significant capital investment to build a physically extensive network that resulted in economies of scale and falling costs. They also provided services considered essential, and their smooth operations were regarded as important for state security.

Electricity supply companies globally in the early and mid–twentieth century typically displayed a virtuous cycle of expansion followed by price cuts followed by more expansion. The growth in consumer demand for electricity provided the income that paid for new generating equipment. New generating equipment tended to be larger and more efficient. As electricity supply companies developed a larger and more diversified base of customers, too, the plants ran closer to their maximum capacity for more of the day, amortizing the capital cost over a larger number of units sold and further lowering the cost per kilowatt-hour. Adding to the incumbents' advantage, the building of electricity transmission and distribution networks was time-consuming and difficult, involving not only significant amounts of capital investments and technical expertise but also the ability to negotiate rights-of-way and other uses of property.

Hong Kong was a global outlier in having no regulatory framework, no franchise, and thus no way of regulating the monopoly that its electricity supply companies exercised in their territories. The only legal or regulatory control on the colony's electricity supply companies was the 1911 Electrical Ordinance, a law that was drafted with input from HKE and placed few controls on the companies. The almost complete absence before 1959 of any mention of this ordinance or any other government regulation in the thousands of pages of primary and secondary documents I reviewed demonstrates that government oversight of the electricity companies before this date was essentially nonexistent. There are only eighteen records relating to "electricity" in the indexed CO 129 record of colonial government correspondence from 1841 to 1951 compared to 856 relating to "water."[6] Perhaps a more telling indication of the lack of legal and administrative oversight over electricity can be found in the November 1959 expiration of CLP's thirty-year agreement allowing it to electrify the New Territories. The ESCC finished its public hearings in the same month. Although attention in Hong Kong was focused on electricity supply as never before, it was two years before anyone in government noticed that the agreement had lapsed.[7]

Following the 1911 Electrical Ordinance, there had been at least four other government interventions, or attempted interventions, in the affairs of the electricity supply companies before 1959. Beginning in 1921 and ending in 1926, the government conducted negotiations with HKE about limiting annual dividend payments to 15 percent of paid-in capital and 15 percent of outstanding loan amounts, less interest. Under this proposal, half of net profit would be used for rate reductions and the other half would be at the company's disposal to pay out as dividends, subject to the 15 percent restriction. In 1926, that proposal was dropped when the government notified HKE, according to its chairman, that "the electricity supply given by the Company [was] good and the charges reasonable and that any form of control by Government was unnecessary" except insofar as safety was concerned.[8]

The next point of intervention came with the contested right to electrify the New Territories. Like CLP, HKE in the late 1920s also wanted to electrify the area. In 1928 Elly Kadoorie and J. P. Braga appealed directly to

ELECTRICITY AS A POLITICAL PROJECT

Governor Cecil Clementi and successfully won the New Territories electrification rights in what marked the first recorded Kadoorie intervention in CLP's operations. The fact that the two companies sought rights from the government suggests that there was an implicit concept of an exclusive franchise. The next year, in 1929, CLP's general manager, Shewan Tomes, signed a thirty-year agreement with the government's Department of Public Works giving it rights to "lay down, erect and maintain electric light and power cables and lines" to electrify the New Territories. The agreement pledged the government's help in laying wires and resisting "extortionate or unreasonable demands" from property owners and mandated that no competitor could receive more favorable terms from the government. There were only a few service standards specified in this agreement. The company had to supply 50-cycle electricity (the number of oscillations in a minute that alternating current (AC) undergoes, typically now either 50 or 60 Hertz) and the frequency could not vary by more than 2.5 percent. At the end of the thirty years, the company would be responsible for removing all wires, cables, and poles. Although this agreement granted CLP the right to supply electricity under certain conditions, it was not an exclusive franchise. Competitors could not be given more favorable terms than CLP but were otherwise free to enter the territory. Unlike most franchises, no royalties were payable.[9]

The third and fourth instances of intervention, as detailed in chapters 3 and 4, came after the end of World War II. The electricity supply companies were operated by the British military from September 1945 until June 1946, and HKE's and CLP's owners negotiated tariffs and other conditions with the military administration and then the civilian government before the operations were returned to private management. Then, in 1948, the government attempted to impose price and financial controls on both companies. Although CLP and HKE retained an unusually high degree of corporate autonomy compared with their global peers, it is important to note the historical antecedents in the 1920s and 1940s that presaged the more sustained campaign to successfully establish oversight over the electricity supply companies in the late 1950s and early 1960s.

In Britain, electricity distribution had been subject to stringent restrictions from its inception in the 1880s, thanks to the Electric Lighting Acts of 1882 and 1888.[10] When the state did not act quickly enough, electricity users pressured it to control prices and often to dictate conditions of service.

This was more typical in the United States.[11] Utility executives in Europe and the United States accommodated to these demands for state oversight. Samuel Insull, earlier one of Thomas Edison's most trusted assistants and later the most powerful utility executive in the United States, warned as early as 1898 that regulation was inevitable. By accommodating themselves to state control, Insull maintained that electricity supply companies would benefit from a sense of certainty that in turn would bolster continued growth. He knew, as Kadoorie was to discover six decades later, that "the one supreme advantage of long-term franchise security was that it permitted long-term low-cost financing."[12] Insull also outflanked the institutions that were intended to regulate him. He supported regulation by state utility commissions, which, as Insull correctly foresaw, would be less corrupt than the municipal commissioners he had dealt with in Chicago, but not as powerful or wide in vision as federal regulators might be.[13]

Electricity pricing was one issue that led to calls for state controls. Availability was another. Electricity supply companies typically concentrated their service in dense urban areas and resisted calls to serve rural customers. This was particularly pronounced in the United States with its large land mass and dispersed population. Both the Tennessee Valley Authority and the Rural Electrification Administration were government-created bodies that attempted to fill this gap in the face of determined resistance from private electricity supply companies.[14] Robert Caro details the stubborn reluctance of electricity supply companies in Texas and elsewhere to provide electricity to rural areas. Without pumped water or electric light or refrigeration, farm productivity remained depressed and farmers' lives unnecessarily primitive.[15] In Hong Kong, both cheaper electricity prices and expanded availability, especially for rural New Territories residents, would be important points of contention in the ESCC hearings.

State oversight of electricity supply was the norm in other British colonies in Asia. India consolidated legislation with the Electricity Act of 1903 and substantially amended this in 1910 before transferring electricity oversight to provincial governments in 1919 and nationalizing most electricity supply companies after independence in 1947.[16] Singapore and Malaya imposed legislative controls on electricity supply and prices.[17] It remains a mystery why state oversight of electricity in Hong Kong was negligible. To recap: oversight of HKE was almost entirely absent, the 1911 Electrical Ordinance notwithstanding; CLP appears to have had no official agreement except that governing the New Territories, where Governor Clementi's involvement seems primarily to reflect the need to settle a commercial dispute between CLP and HKE over which company would secure first-mover rights for the area. Although hard evidence is lacking, the most obvious explanation is that Hong Kong administrators generally had a hands-off attitude toward business, and this policy extended to electricity until the post-1945 period.

What changed in Hong Kong between 1948, when regulation of the electricity supply companies was abandoned, and 1959, when the idea was embraced with renewed vigor? Part of the answer was that getting the public to focus on electricity rather than other difficult political and social issues provided the government with, as mentioned above, a "sort of safety valve." But for electricity to fulfill that function meant that it had to be regarded in a new way. No longer was electricity thought of as an expensive and novel technology destined for the rich alone. Before World War II, electricity in Hong Kong had been a luxury good; the governor in the 1920s was personally responsible for paying his electricity bill. CLP's confusing, opaque tariffs stoked public unhappiness. Factory owners paid less per kilowatt hour to run their machinery than apartment dwellers in Kowloon paid for their lights. New Territories villagers paid more for lights than did Kowloon residents. All had to pay a widely disliked and little-understood fuel surcharge.

During the 1950s, electricity supply was on its way to being regarded as a fundamental right, one whose supply business owners and ordinary people increasingly took for granted as essential to life in the colony. Electricity was a key part of what defined the midcentury metropolis. For those who had not yet thought to ask for electricity, such as New Territories villagers, the government helped articulate a desire for current. Increasing consumer and citizen expectations meant that the electricity companies were in some sense penalized by the success of the service that they provided.[18] The democratization of electricity thus made it politically acceptable to hold the electricity supply companies more accountable for their pricing and service.

The expanded and more powerful colonial administrative government was asked and expected to do more to provide for people's needs and to regulate and control increasingly complex systems. After the failures of the 1930s Depression, governments and their central banks were determined to engineer macroeconomic stability. Even governments that claimed to be hands-off found themselves adopting more interventionist policies. By the standards of the late nineteenth or even the early twenty-first century, Financial Secretary John Cowperthwaite was economically interventionist and had limited regard for property rights.

Hong Kong was not laissez-faire. It was simply less interventionist, especially in its trade policies, than its neighbors, such as South Korea, Taiwan, the Philippines, Malaysia, and Indonesia. Most observers have examined only one side of the issue, pointing out that Hong Kong had freer policies and lower taxes than most countries. But there is no denying that Cowperthwaite saw government action as necessary in some cases to ensure economically rational policies. Far from being a Friedmanite market fundamentalist who thought that companies need act only in the interests of their shareholders, Cowperthwaite often expressed disdain for "speculators" who traded shares; he evinced little concern for shareholders' rights when they stood in the way of his policy objectives.

What we see in Hong Kong is the development, beginning in the 1950s, of dramatically increased administrative intervention. Even before electricity was targeted, state oversight had already been extended to other industries that were considered natural monopolies. In 1951, the Star Ferry Co. had agreed to a fifteen-year franchise, with monthly royalty payments of about 25 percent of net profit. In 1953, the other ferry company, Hongkong & Yaumati Ferry Co., signed a similar agreement, retroactive to 1950. In 1925, Hong Kong Telephone had entered into a fifty-year franchise, subject to royalty payments; this was revised in 1955.[19]

The electricity supply companies worked with colonial administrators to bring clean, safe light where there had been dirty and dangerous kerosene and candles.[20] Electricity was modern. Electricity helped bring order out of chaos. Meters regulated consumption. Numbered electricity transmission poles and maps of underground and overground transmission and distribution networks allowed the electricity companies to regulate themselves. The maps and drawings allowed the company and the state to understand and control more of the colony's territory and its people. The electricity companies laid down and imposed a new form of order on city life. This was part of a larger process of order and control imposed by the modern state.[21]

THE FORMATION OF THE ELECTRICITY SUPPLY COMPANIES COMMISSION

Financial Secretary Arthur Clarke sent Kadoorie a letter in April 1956, with no apparent warning, calling for all the colony's monopoly or semimonopoly enterprises serving the public to be nationalized, with the government "ultimately obtaining a controlling interest in all the utilities." For Clarke, this list included not only the electricity companies, but the telephone company, the bus companies, the gas company, and the Star Ferry company. Reflecting his conviction that the government would have its way, Clarke added: "I note that some companies including, I believe, the Star Ferry, have provisions limiting the number of shares that can be held by any one person, but we should have to get around that in some way or other." Kadoorie's emphatic rejection of the finance secretary's proposal came in the form of a four-page letter two weeks later, a blistering twenty-nine-point response that concluded: "I feel sure there must be some good points in your suggestion but I am unable to appreciate them without further and more detailed information."[22]

At the heart of the dispute were two competing visions of CLP and of the role of an electricity company in a mid–twentieth century city. Clarke, his successor John Cowperthwaite, and other critics saw CLP as a company that used its monopoly pricing power to charge rates that were higher than they would have been had there been competition. Critics charged that CLP's power charges for large users, such as factories, were significantly higher than in Singapore or Tokyo. Moreover, critics contended that they used these excess profits to pay for increased generating capacity, capacity which in turn generated more profits for the Kadoories and other shareholders using funds that Cowperthwaite vigorously argued should belong to the consumers who had provided them.

Kadoorie and those inside the company saw CLP as a firm that struggled to find customers in a fair, free, and intensely competitive market. Theirs was a world where consumers had choices. "Not a monopoly," Kadoorie wrote in the margin of a statement by CLP critic Charles Barber. CLP, Kadoorie noted, promised customers reliable supplies of electricity at a reasonable price. It invested substantial amounts of shareholders' funds to build new generating capacity in anticipation of future demand. Customers had choices; they were free to put in their own generators, or to locate outside of the CLP service territory if they didn't like CLP's terms. Hong Kong had two electricity companies, offering unusual competition. "Anyone can start a new electric company. There is no monopoly," he wrote.[23] Indeed, some of the larger companies in the machine industry had their own generators as late as the mid-1950s.[24]

Although HKE also appeared at the ESCC hearings, CLP—represented by Lawrence Kadoorie and others—was the primary target both during the hearings and in subsequent negotiations. CLP was the larger, more profitable, and faster-growing electricity company and the supplier to most of the colony's factories. The company was particularly aggressive in its use of higher rates to fund expansion. Were other factors at play? Leo Goodstadt, a former colonial official who authored a number of scholarly works on the territory's administration, says that the campaign against CLP was being waged by Shanghai elites. There was also the sense that rationalizing electricity supply by merging and nationalizing companies was, as the government had expressed it in 1948, a "trend of the times." CLP's majority Jewish board and the Kadoories' almost 40 percent ownership stake in the company also made it easier to go after that firm than the more conventionally British HKE, which was controlled by various locally incorporated British *hongs*, or business groups, including Hutchison.[25] Moreover, electricity's growing importance in everyday economic life, and the price that businesses had to pay for it, fueled a sense that it was too important to be left in the hands of a private company.

In his 1956 letter arguing in favor of nationalization, Clarke refers to a letter that Kadoorie had written to Governor Grantham in August 1955. Neither Kadoorie's letter nor any related correspondence with Grantham has been found, but its mention provides yet more evidence that the ownership and control of electricity supply and other large-scale projects were issues in which the government maintained an ongoing interest. It was an interest that surfaced intermittently, in bursts, after which it subsided. For two and a half years after Clarke and Kadoorie exchanged letters, the historical record regarding government control over the electricity supply companies is blank. Then, in late 1958, a campaign against CLP started in earnest. Spearheaded by the *Hongkong Tiger Standard*, with active support from the Reform Club of Hong Kong and some of CLP's largest industrial power users, its proponents called for cheaper and more reliable electricity, espousing the cause of helping Hong Kong's economy, especially its manufacturers.

An editorial on November 30, 1958 in the *Hongkong Tiger Standard*, titled "Hongkong's Power Problems," marked the start of the campaign. The editorial took as its starting point reports that nuclear power was being studied for its possible use in Hong Kong (a fact that Kadoorie had recently mentioned in his 1958 shareholder's letter) as the reason to have "an exhaustive and impartial survey into our power problems."[26]

The *Standard* counseled that it was better to "keep our feet firmly on the ground and look into immediately feasible means of helping our growing industries with cheaper and better power supplies" rather than rush into a new technology with high initial costs. The paper called for a commission to examine the colony's power needs, noting that as a "growing industrial city" Hong Kong's electricity charges needed to be competitive with others in the region. The paper also questioned whether a colony as small as Hong Kong needed two separate electricity companies, "with their separate personnel and generating facilities." On December 5, the *Standard* returned to the topic in an editorial arguing that power rates were among the most urgent topics facing industry. Subsequent articles over the next week noted that Hong Kong's bulk power rates of 95 Hong Kong cents per kWh compared unfavorably with those in Singapore (68 cents) and Japan (77 cents).[27]

On December 9, 1958, the *Standard* ran a news article entitled "Power Breaks Halt Operations of HK Plants." The paper, citing data from an unnamed source, reported that power stoppages had forced factories in what was termed the New Territories industrial complex to halt operations for 3,354 minutes during the two-year period ended June 1958. (The accounts are ambiguous as to the location, but it was likely the Kwun Tong industrial complex or the Cheung Sha Wan factory estate.) This would mean nearly fifty-six hours of stoppages over two years, or a little more than one hour every other week. That shortages of this duration would elicit such criticism reflected the newfound expectation that electricity supply should be uninterrupted.[28] The paper added that the power cuts had resulted in production losses and hurt the wages of workers paid by piecework. (There was no suggestion that the power cuts damaged equipment or required the factory to engage in time-consuming processes to restart production.) The next day, the paper ran an article, "Civic Leaders, Industrialists Express Concern Over Power Breaks," that highlighted the issue of intermittent supplies and the continued fuel surcharge, which had been introduced in 1951 and was intended to pass along the cost of fuel hikes.

The role of the *Standard* in this campaign is worth examining. The *Hongkong Tiger Standard* began publication in 1949. It was launched by Tiger Balm muscle ointment founder Aw Boon Haw, a successful Burmese-Chinese businessman and KMT supporter, who had earlier started the Chinese-language chain of *Sing Tao* newspapers. The *Standard* was dedicated to providing an Asian voice, particularly by standing up for Chinese interests in colonial Hong Kong. It was owned and staffed largely by Chinese. Its first editor, L. Z. Yuan, was a highly regarded U.S.-trained journalist.

The *Standard*'s senior Chinese editorial staff and crusading spirit stood in contrast to the pro-British establishment *South China Morning Post*, whose owners, and most of its editors and subeditors, were British or Australian.[29] The *SCMP*, writes John Lent in a study of Asian newspapers, was "generally regarded as an unofficial spokesman for the government" and at the same time "tend[ed] to be more favorable to Red China." The *Standard* had a "pro-Nationalist and pro-American" slant and its editorials were "much more critical of the government and of Red China." Its circulation of around 11,000 and a readership of some 20,000 trailed those of the *Post*, which in the 1960s claimed circulation of 25,000 and readership of 70,000.[30]

By the late 1950s the *Standard* had carved out a significant place in Hong Kong's English-language media as a young, crusading paper. Along with the civic and political reformers and recently arrived Chinese businessmen who were its sources and, presumably, a significant part of its audience, the *Standard* was ready and willing to attack the old ways of doing business.[31] Among those quoted on the electricity issue were Brook Bernacchi, chairman of the Reform Club of Hong Kong, and C. C. Chu, of the Hong Kong Cotton Spinners Association, whose members were among CLP's largest customers. An unnamed "director of the boards of a number of companies" asserted that funds for new power plants should not come from consumers but from "fresh capital." One unidentified industrialist said that he was thinking of organizing a committee and seeking legal advice to see if damages were warranted for the power interruptions. The industrialists interviewed were also "unanimous" in declaring that the fuel surcharge should be removed.[32]

A December 11 *Standard* editorial, "More Power to You," accused the government and the power companies of behaving like ostriches. The paper warned that this "'bury-your-head-and-the-unpleasantness-will-go-away' approach is useless in solving problems." The *China Mail* joined in, questioning the need for the surcharge at a time of high profits, with an article entitled simply "The Surcharge."[33]

The campaign against CLP was part of a broader push by business leaders for government subsidies for industry. Sociologist Wong Siu-lun notes that although businessmen supported laissez-faire policies in general, they supported government intervention on specific issues. The 1959–1960 period saw industrialists lobby for the formation of an industrial bank to fund business. Cowperthwaite was appointed on January 1, 1959, to chair a committee to examine the establishment of such a bank. In June 1960 the committee recommended against establishing the new bank.[34]

Management at CLP and HKE took the articles seriously enough to discuss a formal response. CLP director Lo Man-kam (usually known in English as M. K. Lo) drafted a response, but in the end Kadoorie convinced his board not to respond. He believed that the controversy was grandstanding by two competing organizations, the Civic Association and the Reform Club, and that the furor would die down.[35] Kadoorie's notes from a meeting with senior HKE executives (Chairman G. T. Tagg, General Manager Stoker, and Flanagan) on December 16, 1958, report: "It was felt that the present agitation had been supported by the Civil [sic] Association on insufficient evidence—to jump on a popular bandwagon before their rival, the Reform Club, could do so." Kadoorie justified the electricity supply companies' lack of response by arguing: "To excuse is to accuse."

In what Kadoorie thought was the unlikely event that the government went ahead with a commission to examine electricity supply, he felt sure that the companies, working through Unofficial (i.e., appointed nongovernment) Legislative Council members, would be able to influence the terms of reference.[36]

This cavalier dismissal of the campaign was to be a serious misjudgment on Kadoorie's part. His attitude toward the anti-CLP campaign reflected his tendency, noted by Leo Goodstadt, to believe that only the elites mattered.[37] Despite his many philanthropic activities, ability to build broad global networks, and a prescient view of many future trends, Kadoorie displayed little understanding of the grievances held by a more demanding generation of Hong Kong Chinese and their expatriate allies. This blind spot was a legacy of the elite world of colonial Hong Kong and Shanghai's International Settlement where he had spent so much of his life. His disparagement of the initial call by the Reform Club as little more than an attention-grabbing stunt to compete with the Civic Association fits this pattern of viewing Hong Kong through an elite lens. It was also a telling strategic miscalculation indicating that he had failed to register the beginning of a more activist political climate in Hong Kong.

Kadoorie may genuinely have believed that the Reform Club of Hong Kong was competing with the Hong Kong Civic Association, headed by Hilton Cheong-leen. In fact, the two organizations complemented one another, and Bernacchi and Cheong-leen were personal friends.[38] The two groups, along with the Chinese Reform Association, headed for many years by Percy Chen, were not just grandstanding. They constituted the closest thing that Hong Kong had to opposition political parties in the postwar period. If they were competing, it was to find issues that would resonate with a politically and socially more adventurous Hong Kong population.

The Reform Club's Bernacchi, a Cambridge and Westminster School graduate, was a leading barrister and one of the most forceful voices for political reform in Hong Kong in the 1950s. Bernacchi formed the Reform Club to support the reforms proposed by Governor Mark Young and was an inveterate voice in favor of popular elections in Hong Kong. He held an Urban Council seat for all but five years from 1952 to 1995.

The Reform Club had a lengthy agenda that included tackling corruption, easing the housing shortage, removing restrictions on rice imports, reducing unemployment, improving health and schools, rent control, marriage law reform, use of the Chinese language, and encouraging emigration from Hong Kong.[39] Bernacchi and his Reform Club were part of a growing constellation of individuals and groups advocating for change. Hilton Cheong-leen was another reformer. Born in British Guyana, he was a cofounder of the Civic Association in 1954. Elected to the Urban Council in 1957, Cheong-leen was a key figure in pushing for compulsory primary school education, which Hong Kong adopted in 1971. He was also vice chairman of the United Nations Association of Hong Kong, chaired by his friend Ma Man-fai. Ma had also been a member of the Reform Club for a time, before separating from Bernacchi. He also emerged as a critic of

ELECTRICITY AS A POLITICAL PROJECT

CLP during the ESCC hearings.[40] Reformers had a long list of proposed improvements. In the late 1950s and early 1960s, they turned a critical eye on CLP's operations.

Four months later, on April 9, 1959, the government announced the formation of the ESCC. The official announcement in the *Daily Information Bulletin* stated:

> Government has given long and careful consideration to representations which have been made during the past few years regarding the position of the two electricity supply companies... who exercise effective monopolies but who, unlike other public utilities in Hong Kong, are subject to no form of Governmental control, particularly in respect of their charges to the public. Government has now come to the conclusion that it is undesirable that the operations of these two electricity supply companies should remain entirely free from any statutory control.[41]

Despite Kadoorie's repeated assertions that the electricity supply companies would be able to dictate the direction of any inquiry, the companies were not consulted before the government's announcement of the ESCC's formation. Each company's shares fell a little more than 3 percent in response to the news in the following day's trading. CLP shares dropped from $17.90 to $17.30. HKE's shares fell from $28.10 to $27.20.[42]

The government's *Daily Information Bulletin*, issued by the Public Relations Office, placed the ESCC news second in its daily summary, behind the announcement that the Medical and Health Department on April 13 would start its annual anti-typhoid campaign, featuring free vaccine injections. The previous year had seen 816 reported typhoid cases, and thirty-four deaths, with almost half the cases occurring in people under fifteen years old. The priority given to news of the typhoid vaccine underscored the variety of challenges Hong Kong faced.

The *SCMP* editorialized: "Everywhere now the importance of cheap power is accepted, though power is essential even if it is not cheap." The paper noted that the presence of "two large companies in this small Colony still leaves them not competitors but monopolists in their separate spheres." The newspaper approved of the fact that outside experts would be brought in to help rationalize the colony's electricity situation, noting that in Britain "central planning and regional amalgamations transformed the whole power situation just in time to cope with the advent of atomic power."[43] The *SCMP*'s editorial served as a reminder that the hopes and fears about nuclear power formed part of the context in which the future of Hong Kong's electricity supply companies was discussed. The significant capital expenditures that nuclear power would require demanded better planning and suggested the need for amalgamating the colony's two electricity supply companies.

The two companies tried to prepare for the hearings as best they could. Kadoorie and other executives asked the government to ensure that one ESCC member came from Britain's Central Electricity Board.[44] He expected that such an expert would, given the Hong Kong companies' good service record, shield CLP and HKE from serious criticism. Still, he knew that CLP would likely be attacked on the grounds that it was subsidizing industry.

Kadoorie's worry was flawed, given that industry was leading the campaign against CLP, though he proved prescient in this regard. His initial muted response reflected his belief that CLP was serving the colony, providing ever-cheaper electricity that was powering the factories that were creating the jobs that were helping keep the territory relatively free of political and social turmoil. He viewed the challenge as one of public relations: "We have not quite achieved the reputation of Con Edison in New York of being the 'Company nobody loves' but, according to our critics, our public relations have been 'lousy.'"[45] He understood how electricity's emergence as a political issue could pose a danger to CLP's very existence as an independent company. Yet he appeared to have been unable to comprehend how society was changing and how new social, political, and economic forces threatened the electricity supply companies.[46]

THE ELECTRICITY COMMISSION'S MEMBERSHIP AND PREPARATIONS

Three members were appointed to the ESCC.[47] Two were from Britain. John Mould had recently retired as chairman of the East Midlands Electricity Board. Charles J. M. Bennett was a partner in the accounting firm Barton, Mayhew & Co. The third member, Dhun Jehangir Ruttonjee, was a Legislative Council (Legco) member from 1953 to 1958 and a prominent member of the Hong Kong Parsi business community. His father Jehangir was the first Indian Legco member and founder of the colony's eponymous tuberculosis sanatorium; both father and son had been interned in the Stanley camp during the Japanese occupation. Dhun Ruttonjee, who was knighted in 1957, is strikingly absent from the detailed ESCC budget records; he appears to have received no remuneration or any other support. Mould, the ESCC chairman, was the electricity board expert that Kadoorie wanted. He had earlier been general manager of the Leicester Electricity Department. After the nationalization of Britain's electricity supply companies in 1948, Mould became deputy chairman and then chairman of the East Midlands Electricity Board, one of twelve regional electricity boards, a position from which he had retired just three months earlier, in May 1959.[48]

The government devoted considerable resources and expenditure to this commission. On June 24, 1959, Legco's finance committee budgeted $140,000 for a three-month inquiry. The committee also approved the hiring of two confidential assistants, two clerks, and one messenger. David Whitelegge, the commission's secretary and the most senior colonial official involved in the inquiry, busied himself both with practical arrangements and with ensuring that the ESCC heard from as many stakeholders as possible. Whitelegge arranged for passage and hotel accommodations from England for the two commission members and their wives. Mould received an honorarium of $6,000 a month for the four months he was in the colony; Bennett's firm received $8,400 a month for his services. The government also paid fares for Mrs. Mould and Mrs. Bennett. Each couple had a car and driver at its disposal for the four months they were in the colony, and each occupied an air-conditioned two-bedroom suite in the Gloucester Hotel. The total cost of the commission was $149,375.12, with most of the funds spent on Mould and Bennett's remuneration, accommodation, and passage for the two men and their wives.[49]

Whitelegge arranged for the publication in the *SCMP* of a press release on the expected arrival of Mould and Bennett in mid-July; on August 1, the commission's terms of reference were advertised in the *SCMP*, the *HKTS*, and three Chinese newspapers. Whitelegge wrote to the governor's aide-de-camp to inform him of the arrival of the two British members of the commission in case it should be appropriate to invite them to any events.[50]

Whitelegge apparently recorded the proceedings. He worked with an official at the Hong Kong Telephone Co. to lease two telephone lines connecting the Legislative Council Chamber at Central Government Offices to the Radio Hong Kong Control Room at Mercury House; he also communicated with Radio Hong Kong's controller of broadcasting to discuss arrangements for the two phone lines and to ensure the microphones were properly tested before the hearings started. It is not known if the proceedings were broadcast on RTHK, the colony's public broadcaster. Despite Whitelegge's efforts to record the proceedings, no recording seems to have survived in the RTHK archives.

Whitelegge appeared eager to ensure that critics of the electricity supply companies had a chance to testify before the ESCC. On August 13, Charles Barber, the managing director of Far East American Enterprises Ltd., located on the first floor of 79a Kimberly Road, Kowloon, wrote Whitelegge asking for a thirty-minute interview in order "to discuss certain matters in connection with the Inquiry and the need or desirability of certain exhibits." Barber had been hired a week earlier by four movie studios and was emerging as one of the electricity supply companies' most outspoken critics. Whitelegge answered Barber almost immediately, proposing a meeting the following Wednesday, August 19.[51]

Mould met with the press on his arrival in Hong Kong on August 23, 1959. "He told reporters he had nothing to say at the moment regarding the inquiry in Hongkong. He had yet to learn the situation. 'But I come with an open mind.'"[52] Mould, soon joined by Bennett, was in Hong Kong for a month before the ESCC started its hearings. We know little of their activities. We do know that Mould visited the New Territories on Friday, September 25, just three days before the commission opened its public hearings, in the company of New Territories District Commissioner D. R. Holmes. Mould suggested to Holmes that in his testimony before the Commission he should focus on power outages. Mould also offered his opinion that CLP should be more forward-thinking in its approach to electrifying the New Territories. Holmes reported to Deputy Colonial Secretary David Trench:

> With regard to the proposal that the company should be asked to explain in detail a series of breakdowns, the Chairman of the Commission suggested to us when he visited the New Territories on 25th September that this might be a good line for me to take. He also told us that whilst a private company might well be expected to offer more resistance to extensive rural electrification than would a nationalized board, he nevertheless thought it not

unreasonable for the China Light & Power company to have plans in being for the gradual extension of supply to more and more villages and rural areas.[53]

Mould would almost certainly have already seen a lengthy letter that Holmes wrote to the Commission on September 19 detailing New Territories' villagers complaints about CLP. Holmes' *Summary of Submissions Made by the Rural Committees and Other Associations and Individuals in the New Territories Together with the District Commissioner's Comments* was the result of consultation beginning in late June with local village representatives as well as with Holmes' superiors, and contained significant and detailed information about electrification, both in the main body of the report and in extensive supporting documentation. The submission illustrates CLP's systematic unwillingness to provide New Territories villages with electricity. Holmes contended that villagers should not be left without access to reasonably priced electricity: "Unless it can be made at least as cheap and easy for our country people to enjoy electricity as it is in the town, comparatively few will be able to receive this benefit," wrote Holmes.[54]

THE CRITICS SPEAK

The ESCC did more than catalyze an unprecedented and far-reaching debate about Hong Kong's electricity supply. It sparked a broader discussion about the role of large businesses and about Hong Kong's small-government, hands-off attitude toward social and economic problems. Its most significant long-term impact was on the administrative state itself. Some government officials wanted the colonial administration to take greater responsibility for those who were not benefiting from economic growth, particularly those in squatter villages, resettlement estates, and New Territories villages. The ESCC hearings, focusing as they did on CLP, deflected attention from government failings and shifted responsibility onto CLP. The political arrangements that were eventually negotiated between Cowperthwaite and Kadoorie resulted in a forced marriage of the sort that neither man would have previously imagined but represented a new form of state/corporate cooperation.

CLP was called on to professionalize and rationalize its own operations, just as its electricity was imposing a new sense of professionalization and rationalization, indeed a new order, on the city itself. Hospitals were using more electrical devices. Medical officials demanded that CLP provide the reliable and constant voltage that the machines, particularly sensitive instruments like x-ray devices, demanded. Voltage fluctuations were an issue for factories too, as they caused unnecessary wear on the machines and burned out light bulbs more quickly. As discussed, CLP had long portrayed itself as a technology leader in the colony. Now its users claimed that the quality of the electricity CLP supplied was not up to the standards that a modern city required.

The ESCC was successful in its attempt to elicit complaints about the electricity supply companies. A more detailed list of complaints is below, but the case against the electricity supply companies can be summed up simply: the two companies, but especially CLP, were high-handed monopolists that disregarded consumers.

The fuel surcharge, first imposed in 1941 and then reintroduced in 1951, was particularly controversial because of what appeared to be arbitrary rates. Director of Public Works E. Wilmot Morgan noted, with a hint of frustration, that the "Company has varied its charges by a 'fuel' surcharge which was introduced on 1st December 1951 on which account 19 percent was added, in April, 1952 this surcharge was increased to 23.1 percent and in May, 1952 decreased to 22 percent."[55] CLP statistician Edgar Laufer testified that the company had waived $9.85 million out of the $33.32 million that it was entitled to during the 1956–1958 period. This revelation not only failed to win applause but only strengthened the feeling that CLP was managing the surcharge account arbitrarily for its own benefit.[56]

Even as the ESCC's two overseas members were about to arrive in Hong Kong, CLP and HKE were engaged in business behavior that appeared, at the very least, extremely casual, if not collusive and anticompetitive. In response to a query from ESCC Secretary Whitelegge about a change in the level of the fuel surcharge, CLP wrote: "Referring to your paragraph 2, we would mention that the original warning announcement of the surcharge published on 11th August 1959 was not approved at a formal Board Meeting of the China Light & Power Co., Ltd., but only informally by telephone, after consultation with The Hong Kong Electric Co., Ltd., hence the absence of any minutes on the subject."[57]

The picture that emerged to many was that of an arbitrary monopolist. Variations in tariffs added to that perception. Large factories paid the lowest rates, sometimes less than the cost of the fuel used to generate electricity. HKE's consumers on Hong Kong Island paid lower rates for lighting than did CLP's in Kowloon. New Territories villagers paid more than anyone, with their lighting rate per kWh almost four times that of power charges for large factory owners in Kowloon. In addition to the fuel surcharge, a range of other charges, from meter rentals to deposits to installation fees, added to the confusion. There was little transparency in pricing. Information about the surcharge and the range of tariffs was not readily available. CLP employees often seemed to have a take-it-or-leave-it attitude toward customers. The director of public works noted that CLP refused to sign a contract for street lighting services or even justify its rates, unlike HKE, which had a formal contract with the government.

Scores of letters debating the colony's electricity supply system appeared in the English-language newspapers.[58] The public debate over the future of these two large enterprises was unprecedented. Complaints, especially about the telephone company, were already a common feature of the letters pages in the colony's newspapers. The ESCC hearings, which were conducted between September 28 and November 17, and their aftermath elevated the correspondence from the personal grumblings of an individual letter writer to a community attempt to articulate and debate the broader policy framework that governed the companies providing public goods and services.[59]

However, many correspondents, if not most, defended the current electricity supply arrangements. They noted how much better the electricity supply companies were than the waterworks and the telephone companies and compared Hong Kong's electricity supply favorably to that in Britain. This sentiment had a factual basis. Economist Edward Szczepanik noted that Hong Kong was "probably the only place in the Far East to have had no postwar power shortage." Indeed, he cited the expansion of electricity production as "the best example of the adaptability of Hong Kong's infrastructure to the demands of industrialization."[60]

Some letter writers saw the struggle for electricity in broader ideological terms, as a Cold War battle between the forces of freedom and communism. One, who voiced similar sentiments at a CLP shareholders' meeting, warned that nationalization would lead to "Red electricity, just as Hong Kong now had 'red water.'"[61]

Conversely, Percy Chen, of the Chinese Reform Association, wrote that CLP was Robin Hood in reverse. "Robin Hood took from the rich to aid the poor. China Light takes from the poor to help the rich. That is a situation which I don't think any self-respecting community can tolerate."[62] Ma Man-fai, a member of the founding family of the prominent Sincere Department Store and cofounder of the Reform Club (thus part of the circle of reformers that included Chen, Bernacchi, and Cheong-leen), expressed similar sentiments. Ma told the *SCMP* that Hong Kong was "an economic jungle" and that the ESCC constituted "a breath of fresh air." Ma, identified in this article as a member of the Po Yick Association (a retail trade group), paid tribute to the commission "for their excellent demonstration of the British way of life and hoped that when its members left Hongkong, the colony would not lapse into its jungle again."[63]

Some critiques were more far-reaching, using the ESCC probe into CLP to mount a broader assault on Hong Kong's economic and political power structure. A correspondent who signed himself or herself "Beatnik" was one of the most searing. Beatnik's letter was published on January 21, 1960, some ten days after the ESCC recommendations that the electricity supply companies be merged and nationalized. The letter explicitly responded to Kadoorie's evocation of the "peculiar" Hong Kong conditions, spirit, and policy. Beatnik tied the sweatshop conditions and widespread poverty in Hong Kong to the monopoly power exercised by enterprises like CLP:

> The long rows of shabby tenements where men live like animals; the dilapidated Victorian facades which are an eyesore on Nathan Road; the hovels, the filth. . . . As for the long range "planning" of the authorities, the result is typical of the method. They still don't know how much water is being used, or will be used. Relief food is sold and carted openly round the corner in endless rows of blackmarket trucks, scooters and bicycles.
>
> All this is the Hongkong "spirit." Call a spade a spade. I think it's the worst form of colonialism. It is their fetish, their Deity. But they needn't worship it so passionately.[64]

Beatnik's letter reflected the changing social and political context of Hong Kong. A new and more numerous group of social reformers took on a newfound importance. In addition to Bernacchi, Chen, Cheong-leen, and Ma, social reformer and former missionary Elsie Elliot (later Elsie Tu) began working in Hong Kong in the 1950s.[65] W. S. Edwards wrote a series of letters to Prime Minister Harold MacMillan questioning the independence of the commission. Edwards's first letter, which includes newspaper clippings dated April 14, 1959, announcing the formation of the ESCC, noted that at least four directors of the electricity companies were either Executive Council or Legislative Council members. A second letter, written May 7, noted that CLP had donated $100,000 to the University of Hong Kong toward a $700,000 fundraising campaign for a steam power laboratory. In the past, Edwards noted, CLP had only donated $10,000 or so a year to the community.[66]

CLP's most visible and tenacious opponent was Charles H. Barber, an American ex-serviceman, who had, by his own account, earlier in his career worked for the municipality of Lansing, Michigan, served in the U.S. military and, in Hong Kong, worked for a time at HKE.[67] Barber was employed as a consultant by four movie studios for the ESCC hearings, yet he appears to have operated largely independently. He had no apparent ties to Shanghai textile interests or to the civil society groups such as the Reform Club. Barber's testimony at the ESCC suggests that his primary influence in envisioning the political structure of electricity supply systems was the municipal power movement in the United States. Although Barber is a singular figure, apparently largely acting alone, it is he who appears as a key protagonist in newspaper accounts and in the public hearings. Tellingly, it is Barber to whom Kadoorie devotes much of his attention, at least as reflected in the Kadoorie archives. Kadoorie's focus on Barber, and his frustration, is reflected in a handwritten annotation: "Barber appears to have set up his own commission—He has all the answers—can we work out what compensation CL&P should get if Barber's suggestions are adopted."[68]

The four movie studios hired Barber as a consultant one week after the government published its notice about the formation of the ESCC. The companies were Asia Pictures Studio, Yung Hwa Motion Pictures Studio, Wadar Motion Picture & Development, and Shaw Studio. They were significant users of electricity and paid the higher rates imposed on electric lighting than on power for industry. Barber's remit seemingly went well beyond his stated mandate to analyze electricity standards and charges compared with other places in the world. He developed a coherent and far-reaching analysis arguing for statutory state control of electricity supply companies. The unnamed spokesman quoted below is almost certainly Barber:

A spokesman for the industry said yesterday that studios believed that extensive statutory control of the electricity supply was essential to provide better service at the lowest possible price. "Government regulation of monopolies is accepted everywhere as a necessary public function. The local electricity supply companies are monopolies, and as such, should be operated in the public interest," he said. "Progressive public regulation has almost everywhere resulted in fair rates and good service for consumers while preserving the principle of a fair rate of return on public utility investments."[69]

Barber, if it was indeed he, set out the case for regulation of monopoly businesses, something he would do in more detail during the hearings. The Chinese Manufacturers Association weighed in two weeks later, and a front-page article in the *SCMP* registered the group's complaints. These included charges that CLP used old-fashioned equipment and thus had more expensive power; that manufacturers deserved compensation for the frequent power stoppages, which curtailed production; that New Territories charges were much higher than those in urban areas; that large consumers paid much less than midsized ones; that the 1 percent annual interest rate paid on deposits was too low; and that some customers were cut off for nonpayment even when the amount due was less than the deposit.[70]

As noted above, New Territories District Commissioner Holmes prepared a lengthy *Summary of Submissions*. Holmes's report, in response to an order by Deputy Colonial Secretary Trench, is by far the most extensive submitted by any government body, longer than those of all other government departments put together. We do not know precisely how Holmes prepared the *Summary of Submissions*, but it was the product of intensive discussion with representatives of New Territories villagers. Holmes shared his conclusions with the New Territories representatives and was at some pains to let them know that they could also make personal representations to the Commission. He also worked with the Shek Wu Hui Chamber of Commerce, which made detailed records of power cuts in Shek Wu Hui and Fanling.[71] Holmes's report reflected concern on the part of at least some government officials that New Territories' villagers were not enjoying the fruits of development in part because they did not have access to electricity.

Holmes, as part of this exercise, also coordinated an exhaustive tabulation of electrification in New Territories villages. The conclusions were stark: Fewer than one in six villages had electricity. In Tai Po District, only twenty-one of 280 villages had supply; in Yuen Long, forty-nine of 162 villages had electricity; in Tsuen Wan, the figure was eighteen of forty-three; in the Southern District, which included Sai Kung, a mere fifteen of 149 villages, just 10 percent, were electrified. In total, only 103 of the New Territories' 634 villages had electricity.[72] CLP's prejudice against rural customers was typical of electricity supply companies in many countries; the cost of building distribution lines in areas with low population density meant that the financial return on serving farm and village customers was relatively unattractive. As noted, in the United States, the creation of the Rural Electrification Administration and the Tennessee Valley Authority were designed to ensure that rural areas were adequately supplied with electricity. Less well-known is Samuel Insull's success in profitably serving rural areas.[73]

New Territories villagers had a long list of grievances. The main complaints centered on the cost and time it took to extend electricity to villages, as well as the conditions that were attached; the unreliability and instability of electricity supply; higher light charges in the New Territories than in Kowloon; high deposit costs and the low interest rate paid on them; the difficulty of making bill payments in person; the fuel surcharge; the poor maintenance of facilities; quick cutoffs of electricity in cases where bills were not paid on time; and the use of private property for rights-of-way without adequate notice or compensation. Some specific examples of these complaints follow.

A Sha Tau Kok representative complained of a $150 per customer installation fee. One person near Sha Tin claimed to have been asked for $4,500 for an electrical connection; further complaints were made about charges for installing larger cables and transformers when the number of consumers increased. Separately, a shop in Texaco Road, Tsuen Wan, "which is only 15 yards from the supply line," was asked to pay $750 as an installation fee. There were complaints about the $500 CLP charged for each electricity pole installed. There was also unhappiness that on occasion additional charges were made for the cost of wiring. "It is stated that no such charges were made when supply was installed in Ha Tsueng and Ki Lun Wai prewar," reported Holmes.

Some villages were forced to accept collective responsibility by agreeing to install a main village meter as well as individual meters and were forced to pay extra when the consumption recorded by the main meter exceeded that of the combined households. Holmes commented, "I am informed that this is only done in areas where evidence has come to light suggesting that electricity is being stolen." Correspondence by some of the company's most senior managers displayed a condescending, combative attitude toward villagers. Station manager Cyril Wood, second only to Kadoorie at the company, spoke of the need to ensure that Tai Loo Ling Village was "well-behaved." Another senior manager, acting station manager Eugene Joffe, spoke disparagingly of "villages of this type" in justifying a large communal deposit and a master meter. Also evident is frustration on the part of CLP managers at the theft of materials, particularly copper wire, as well as electricity.[74]

The issue at Tai Loo Ling centered on CLP's conditions before it would reconnect the village to its electricity supply. CLP had disconnected the village in May 1947 for nonpayment of account. Before reconnecting the village, CLP staff demanded that at least 70 percent of the debtors pay their bills; that CLP be allowed to overhaul and recalibrate all meters; secure a deposit of at least $50 per customer; and install both a master meter for the village and individual meters. The company said that if there was more than a 5 percent discrepancy between the total of the household meters and the main village meter that the difference must be paid for by a surcharge on individual accounts the following month or from a special communal account; the special communal deposit was "to be used as a guarantee that the village is well-behaved, and also for the payment of any differences between the master meter and the villagers' individual meters," station manager Wood wrote on July 25, 1956, to a contractor who was inquiring on the villagers' behalf. In a cover letter forwarding a copy of Wood's letter to the government for the commission to review, Acting Manager Joffe wrote: "Owing to prevalence of theft in villages of this type we have found it necessary to install a master meter and insist upon a communal deposit."[75]

Kadoorie and CLP prided themselves on running a modern and rapidly expanding electricity supply system. The reality, as presented by Holmes, was that the system was more unstable than the company admitted. Throughout Tai Po, and especially in Sheung Shui, there were complaints about the reduction of voltage, allegedly due to overloading of the circuit; representatives from Sheung Shui and Sai Kung complained of frequent supply cuts; Kau Wa New Village in Sheung Wan complained that between 8:00 p.m. and 10:00 p.m. voltage was weak and fluorescent lamps could not function. Holmes attached an appendix from the Shek Wu Hui Chamber of Commerce recording power cuts in Shek Wu Hui and Fanling from January 3, 1959, to July 31, 1959.[76]

Representatives complained not only about unreliable electricity supply, but also about CLP's allegedly inadequate maintenance of distribution and transmission equipment. Holmes noted that a woman was killed at Sha Tau Kok pier when a pole that was corroded at the base fell on her; he added that CLP made a prompt *ex gratia* payment to the woman's family. Additionally, Holmes reported that some electricity poles appear to be "live" and thus dangerous. The district officer noted that he had reported at least one instance of this sort to CLP for further action. Higher light charges in the New Territories also caused resentment. Light charges were 37 cents per kWh in the New Territories compared with 29 cents per kWh in urban areas—28 percent higher. CLP did not make it easy for rural customers to pay their electricity bills. At the village of Tai O, payments could only be made one day every two months; at Sai Kung payments were accepted only during a single three-hour period once a month; at Mui Wo it was a single two-hour period each month. These were among the most egregious cases, but the overall picture was of a company that acted as if its customers should be grateful to enjoy the benefits of electricity.

CLP required customers to put down deposits; these were another source of contention. CLP paid only 1 percent interest on deposits, about half of what savings deposits earned. Consumer deposits had almost quintupled in the previous decade, from $3.3 million in 1948 to $16.1 million in 1958. The deposits, which were growing at more than $3 million a year during this period, gave CLP an inexpensive and reliable source of funding at a time when it had no bank debt or other borrowing. By the end of fiscal 1963, the year before the Scheme of Control was agreed to, CLP held $33.9 million in customer deposits.

Another of Holmes's points, reflecting villagers' concerns, states simply: "Abolish the surcharge." No detail is given. The fact that there is no elaboration suggests that the issue of the fuel surcharge was well known. A variety of sources suggest that the confusing and seemingly capricious charges were a significant irritant to many people.

Holmes's New Territories document is revealing for the distance it shows between the villagers' view of CLP and the company's and Kadoorie's self-perception, at least as refracted through ESCC testimony and documents at the HKHP. Holmes's comments provide an additional perspective. He takes pains to say that he has not verified the particular complaints, but implies that, although the particulars may not be altogether reliable, the general picture is correct. His posture is that of a neutral government official. In the archival record, Holmes gave every appearance of scrupulously reporting villagers' concerns, even when he did not agree with them. Notably, he did not understand how CLP could function without deposits for electricity service; he made it clear to the villagers that he would note their call for the abolition of the deposit requirement but could not endorse it.

The ESCC became a hook on which various parts of government could hang their pleas for more state intervention. G. Graham-Cumming, acting director of medical and health services, complained of wide voltage fluctuations. He noted that the United Kingdom permitted only a 5 percent fluctuation, although "high voltage" and "super voltage" x-ray machines were capable of handling a 10 percent variation. Hong Kong voltage variations, wrote Graham-Cumming, were "frequently in excess of this margin. This seriously interferes with the efficiency of the machines [and] as well as being damaging to them has serious consequences on the treatment of patients. It is impossible to standardise treatment techniques when this happens and some patients may receive dangerous over dosage while others may be seriously under treated."[77]

These specific examples illustrate the tension between the need for more regulation and control that accompanied the introduction of more sophisticated technology such as x-ray machines and the more informal approach toward the electricity supply companies that had long prevailed in Hong Kong. Graham-Cumming strongly recommended regulations to keep the voltage fluctuations within 5 percent. The fire department wanted the government to develop and oversee the enforcement of a standardized electricity code. R. G. Cox, acting chief officer of the Fire Brigade, answered Trench's request for submissions on electricity supply practices by recommending the establishment of a standard code in the installation and use of electricity and electrical appliances "and that enforcement of such code of practice should not be left to private enterprises; although under the general direction of a Chief Government Electrical Inspector, certain powers of supervision may be delegated to Electricity Supply Companies."[78]

The head of resettlements had similar concerns about the quality of electricity and also advocated more government oversight of the electricity supply companies. J. P. Aserappa, acting commissioner for resettlement, wrote that electrical installations become unsafe in a very short period of time; the current mandatory five-year inspection period for electrical installations should be modified to require annual inspections; and voltage fluctuations were a problem. "This excessive fluctuation will lead to rapid wear of electrical equipment in the Cheung Sha Wan Factory and to a very high rate of bulb consumption in these [public housing] estates," he wrote, referring to Wong Tai Sin, Li Cheng Uk, Shek Kip Mei, and Tai Hang Tung Resettlement Estates and the Cheung Sha Wan Resettlement Factory.

The electric supply companies should be required to keep voltage fluctuations within a reasonable limit, Aserappa recommended. He attached an appendix that showed the result of voltage tests at these five locations conducted in July and August. The test lengths ranged from thirteen to twenty-four hours in duration. Most tests recorded fluctuations of 5–10 percent from the stated 200-volt norm, with a high of 221 volts and a low of 165 volts. These two readings were the only ones outside of the 10 percent range, that is, lower than 180 volts or higher than 220 volts.[79] Cable & Wireless also submitted a detailed list of shutdowns that had affected its Kwanti Substation in 1957.[80]

Not all department heads appeared to want to cooperate with the ESCC. On September 2, B. I. Barlow, for the director of commerce and industry, wrote: "With reference to the conversation yesterday afternoon between the undersigned and Mr. Whitelegge, I confirm that I have no information which might be of value to the commission in its deliberations." Whitelegge's action in telephoning the Commerce and Industry Department on the very afternoon when submissions were due suggests that he had been hoping to find support from the agency that oversaw the business sector.[81]

The submissions were evaluated by the Department of Public Works. Hector Forsyth, writing for the director of public works, dismissed many of the complaints. The issue of low voltage for x-ray machines was isolated, one that CLP rectified when a substation was subsequently built in the hospital area. Poor wiring was an issue, no doubt, but one that was outside the commission's scope. Choosing to ignore the test results from resettlement estates presented by Aserappa, Forsyth wrote that, with the exception of the

Cheung Sha Wan factory, "where the matter is being rectified," the problem of low voltage in resettlement estates had not been brought to the attention of the chief architect, the chief of electrical and mechanical engineering, or CLP. The list of shutdowns involving the Kwanti Substation could not be introduced as evidence because the material had been provided by CLP to Cable & Wireless in confidence. Forsyth's silence on the issue of CLP's refusal to negotiate a street lighting contract, especially given that it was raised by a colleague in the Department of Public Works, should be read as support for the criticism leveled against CLP.[82]

The outpouring of complaints revealed that CLP's system was not as technically perfect as Kadoorie and CLP's annual reports would have outsiders believe. Kadoorie repeatedly maintained that CLP would supply current to whomever wanted it in whatever amounts they wanted. Supply would be denied to no one. Yet those statements hid a world of voltage fluctuations and power outages. Holmes's report and other government submissions showed that discriminatory pricing policies and an often arrogant attitude made it difficult for New Territories villagers and some businesses to have access to current.

Kadoorie was concerned with CLP's public image. As far back as his 1938 *General Policy* he had stressed an electricity supply company's need for public support and understanding. By global standards, CLP had done a good job in its rapid expansion program, something that the ESCC confirmed. But a combination of overselling just how well its system functioned, coupled with a high-handed attitude (especially toward household consumers), had alienated customers, the very people who both paid its bills and would decide the political arrangements governing its future.

Kadoorie appeared to be living in a different age. Throughout the ESCC hearings, he resisted the claim that CLP was a monopoly power. He repeatedly scribbled marginalia such as "not a monopoly" on documents citing Barber and other critics. Kadoorie contended that industrial factory owners who did not like CLP's electricity service could simply set up captive power sources. The fact that hundreds of factories opened every year during the 1950s in CLP's service area proved to Kadoorie that the company was doing a good job. He was even more out of touch when it came to individual consumers. To him, they were for the most part of little interest and less importance. Consumer lighting was in Kadoorie's mind not a right but a luxury, one for which consumers should be prepared to pay extra in order to subsidize industry and thus nurture job creation.

Under CLP's business model, high consumer prices contributed to profits and provided the funds for further expansion; bulk industrial users provided stability to the load and ensured that the fast-growing network would be more fully used, thus spreading the per-unit costs more widely. The reality that CLP lost money on these industrial users did not seem to concern Kadoorie. Further expansion, fueled by demand from factories, meant, in Kadoorie's mind, lower per-unit electricity costs as well as social stability and economic prosperity. Monopolies, to Kadoorie, used their privileged position to raise prices. It never occurred to him, at least not in his writings, to consider Cowperthwaite's contention that the electricity supply companies were able to charge more than they would have been able to in a competitive market. Falling prices or no, the electricity supply companies reaped the profits of natural monopolies. This had been widely discussed since Insull's time sixty years earlier, yet the worldly Kadoorie somehow seemed unaware of the debate.

He consistently maintained that the company's policy of subsidizing industry was done for the greater good of the colony and in consultation with political leaders. In his closing remarks before the ESCC, he said, "The Commission is mistaken in stating that the company took upon itself, *ex parte*, the responsibility of giving preferential rates to industrial concerns or, in other words, subsidizing industry. This policy was only adopted after consultation and with the approval of Sir Alexander Grantham and Mr. Arthur Clarke who agreed [in 1954] that the serious labor situation and the absolute need to maintain employment in this over-populated Colony was such as to warrant the course adopted."[83] He went on to warn of "the far-reaching and dangerous consequences" which the CLP board "are convinced would result from such a reversal." He took a typically long historical view, arguing that the need to keep employment high and to guarantee Hong Kong's prosperity was particularly urgent, given the likelihood that the territory would revert to Chinese rule after the expiration of the New Territories lease in 1997: "We have seen what has happened to civilizations built up in Shanghai by European nations; how overnight that civilization has disappeared; how the work of a hundred years is gone."[84] The typically upbeat Kadoorie lets the mask slip a bit, deliberately revealing his fears for the colony's future by looking back to the losses suffered in Shanghai, whether by the Japanese or the Chinese Communists. Looking forward to a future with a more "normal" China, "normal in the sense of prewar," he imagines a Hong Kong that could suddenly be void of industry:

> If the situation changes—China opens up—things become more normal in the sense of prewar, there is no logic in having factories, having industries built up in a place where there are no raw materials.... We have industries built up here today on a very special set of circumstances, unlike the building up of industries in other countries where they are dependent on that country itself. There is no logic for industries to remain. Whether they will, or not, only the future can tell. That is another of the special risks that attach to this country [sic] and to this place.[85]

Kadoorie offered to elaborate on his worries but testified that it was in the best interests of all not to do so "in this form." From a man who had seen the destruction of war came the dark forebodings of a possible apocalypse. Photos in the 1959 annual report conveyed a sense of dread, showing as they did damaged equipment and hinting at a colony where "civilization has disappeared" and "the work of a hundred years is gone." Kadoorie correctly viewed the ESCC as the first battle in what would be a prolonged struggle. In December 1959, a month after the ESCC's hearings ended but before its recommendations had been made public, he warned fellow CLP directors:

> The appointment of the recent Electricity Commission by Government is not, as some would have us believe, a final step. It is just the beginning and must serve as a warning that as a public company owning a Public Utility, we must be prepared to fight not only for our own existence but also for that efficiency that has made us what we are. We must expect unwarranted interference and that the ills of Hong Kong's industries will be visited on our heads. We will be the target for labour unrest and the scapegoat for the inefficiency of others.[86]

He was correct that CLP would have to fight for its existence. The ESCC recommendations and the protracted negotiations that led to the SoCA in 1964 jeopardized Kadoorie control of the family's most valuable asset. Kadoorie was right, too, that the company would experience interference.

He could see the dangers and he understood them well enough to turn the inevitable "unwarranted interference" by the government to the company's advantage. Yet he did not seem to appreciate how Hong Kong had changed and how broader social aspirations limited the room for maneuver by the colony's business and political elite. Nor did he fully appreciate how the very success CLP had in normalizing electricity and more fully integrating it into the colony's everyday life also made the company seem less special and heightened public expectations for accountability and service excellence.

NEITHER NATIONALIZATION NOR MERGER BUT ACCOMMODATION

On December 17, a month after the ESCC hearings ended, the commission submitted its report to the government. The recommendations were made public a month later, on January 20, 1960. The ESCC determined that the very idea of a privately owned electricity company was anathema, given the inherent conflict between the interests of consumers and those of private shareholders. The solution was nationalization and merger of the two electricity supply companies. The commission recommended that the government pay shareholders $220 million for CLP and $208 million for HKE in the form of thirty-five-year bonds bearing an interest rate of 6.5 percent annually. At the same time the government warned the companies that following the recommendations of the commission, the government needed to safeguard the assets of the company and prevent them from being stripped in the period before nationalization. Dividends were frozen at the level paid out or forecast by April 10, 1959, when the formation of the ESCC was announced, with the threat that directors would be personally liable for any dividend increase in the event that action was taken to nationalize the assets.[87]

In March 1960, the government announced that the electricity companies had made an "informal approach" to authorities offering to try to meet most of the conditions laid down by the ESCC "short of nationalization."[88] CLP and HKE duly negotiated a proposed merger of their operations to address the complaint of duplicative and inefficient operations owing to the existence of two electricity supply companies in one city. Neither company was enthusiastic, and the government rejected the first plan in January 1961.[89] An agreement acceptable to the government was not reached until May 1962, two and a half years after the ESCC hearings. Its terms were condemned by many CLP shareholders for undervaluing the company's growth prospects. At an Extraordinary General Meeting to consider the change, Kadoorie, as CLP chairman, took the unusual step of adjourning the meeting without calling a vote. He maintained that, based on the proxies he had received, it would be impossible to obtain the needed 90 percent supermajority necessary to approve the merger.[90]

Certainly, the duplication involved in having two geographically distinct monopoly electricity supply companies, each with its own generating facilities and layers of administrative staff, appeared irrational and wasteful to many people in the colony. Editorialists had previously called for amalgamation. The *Hongkong Tiger Standard* had been among those which, even before the ESCC hearings had started, advocated a "thorough and independent inquiry into the company's rates, service, and its status as the sole supplier of power in Kowloon and the New Territories."[91] The *SCMP* had opined that "the fact that there are two large companies in this small Colony still leaves them not competitors but monopolists in their separate spheres."[92]

Cowperthwaite was determined to unify the companies. He was an economic rationalist. He was offended by the duplicate systems of administration, the excess reserve capacity needed because the two companies did not share supply, and the unfair cross-subsidies that saw Kowloon and New Territories farmers and workers subsidizing factory owners. Kowloon factory owners on occasion paid less than the cost of fuel for their electricity. This is consistent with Kadoorie's focus on overall sales and his concern with helping industry rather than protecting residential consumers or ensuring that each class of customer paid its fair share.

Cowperthwaite above all wanted to stop CLP's practice of using revenues to pay for new electricity supply equipment. He was determined to tame these examples of monopoly power and to introduce more fairness in the system, so that a residential consumer in Kowloon or the New Territories would pay the same rates as one in Hong Kong. CLP charged more because it needed funds for its expansion. Cowperthwaite was adamant that this should not continue.[93]

There were intensive and detailed merger negotiations, primarily between the two companies but often including Cowperthwaite. Kadoorie was prepared to accept nationalization provided that the government paid cash or

ELECTRICITY AS A POLITICAL PROJECT

securities worth not less than the issue value of the shares.[94] The companies proposed an arrangement in October 1960 that would have given them a fixed rate of $0.025 of profit and $0.022 for new capital investment per kWh sold. The government rejected the proposal the following January.[95] There were deep divisions over how to value each firm for the purposes of a merger. Although both were electricity supply companies, they had two different philosophies. HKE only expanded when it felt assured that demand would be sufficient to quickly employ the new capacity. CLP expanded in advance of demand and, by the time Kadoorie testified at the ESCC in October 1959, had invested $160 million in the fourteen years since the end of World War II.[96]

The two companies and Cowperthwaite spent much of 1961 attempting to negotiate a deal. The companies, with their separate and quite distinct identities, were never eager to merge. But with nationalization looming as a threat, they needed to be seen to be cooperating with the government. Cowperthwaite drafted an agreement dated November 30, 1961, that proposed "an arrangement not involving compulsory acquisition" but for "a closer integration of the operations of the two companies" and "secondly for the regulation of consumer rates, dividends and related matters with a view to an equitable division between shareholders and consumers of the benefits of future development." He stipulated that the electricity supply "companies will take all reasonable steps . . . to assimilate their operations with a view to facilitating future amalgamation should that be considered desirable." Cowperthwaite proposed very specific interventions in the companies' affairs. They would connect their two power stations "by four 66 kV [kilovolt] cables (operating initially at 33 kV), thus providing for the exchange of 80 MW of electricity." Later, he demanded that the government have the right to appoint two directors to the merged company's board of directors, as well as the first chairman.[97]

Cowperthwaite was similarly interventionist when it came to finance. On November 30, 1961, in the draft agreement he had prepared for negotiations that he intended would lead to the merger of the two companies, he proposed that the companies should raise capital for expansion by "the issues of shares for cash on such terms and conditions as the Governor may approve." Retained earnings, which the companies used to fund expansions, were only to be used as a last resort. He proposed that these be put in a special capital extensions fund. He also demanded that the companies not establish "any new reserves or similar accounts without the approval of the Governor."[98]

Cowperthwaite's proposal likely reflected unhappiness with CLP's establishment of a number of special funds in the 1950s that effectively reduced profits while allowing the company to build up cash reserves. These included a reserve for plant replacement in addition to normal depreciation charges, which themselves are intended to reflect the use of equipment (but can be abused, as in the immediate write-off of one-third of the new Hok Un plant in 1957); a reserve for investment fluctuations; and an insurance reserve. Cowperthwaite also proposed that "in the event of compulsory acquisition by Government at a future date compensation will not be related to the value of capital assets, except to the extent that may be mutually agreed by the Companies and by Government." This was another example of Cowperthwaite's cavalier attitude toward property rights. He proposed that dividends would be permitted to increase formulaically (that is, without regard for underlying financial performance), an additional five cents per share each year, from $1.30 in 1962 to $1.55 in 1967.[99]

In the absence of a bond market, no significant credit from equipment suppliers, and no long-term bank debt, CLP's policy of investing in advance of demand meant that it had little choice but to rely on continuing infusions of new capital from its shareholders. Expansion thus needed to be paid for with the profits from ongoing operations.

The dividends of $1.70 a share paid out in 1959 translated to a yield of around 6 percent for those who bought shares on the open market based on a typical share price that year. The dividend yield was around 15 percent of the cash investment for those who had bought issues of new stock at par value. Those shares were first offered to existing shareholders; the Kadoories made it a practice to take up all of their rights in these share offerings. CLP's financial engineering in fact meant that the Kadoories and other long-term shareholders enjoyed a far higher return on the capital they had invested, with a return on the cash they had put in the business of more than 30 percent annually.[100]

The company had revalued assets in 1951 and given every shareholder an additional share for every share owned. At a stroke, this increased paid-in capital from $21 million to $42 million, even though no additional cash had been paid to the company. An investor who had paid $500 for one hundred shares now had two hundred shares with a par value of $1,000. Par value constituted an asset in the event the company went out of business. Total dividends paid in 1959 were $13.8 million, more than double the 1951 level. During these eight years, the company had raised only $4 million in new capital from its shareholders.

Another bit of financial engineering occurred in 1959 with the transfer of $2 million from the reserve for plant replacement and $53 million from retained earnings in order to increase the par value of shares from $5 to $10 each without requiring shareholders to pay more money. The total paid-in capital of $110.4 million recorded on the balance sheet at the end of fiscal 1959 included $76 million from these two transactions. The Kadoories and other long-time shareholders had invested less than thirty-one cents for every dollar of paid-in capital. That meant the cash-on-cash return, which measures the cash long-term shareholders like the Kadoories received as a percentage of the cash they had actually invested, was 40 percent in 1959. Even the dividend restrictions imposed in 1960, which reduced the payout to 1958 levels, allowed a generous 35 percent annual cash-on-cash return.

The lack of substantial new shareholder capital incensed Cowperthwaite, who believed that shareholders were improperly using consumers' funds to finance expansion: "The money put into the company, apart from that subscribed, did not belong to the shareholders," he contended. "It belonged to the consumers, and no part of this could be taken into consideration, except for a small part being allowed for management."[101] Any new plant and equipment should not be financed using retained funds, which Cowperthwaite said should be returned to consumers. In a letter to Kadoorie, he quotes the ESCC's statement: "All moneys paid by the consumers and invested in the business should inure to the benefit of the consumer." He adds, "Government has accepted the Commission's view on this point, within the limits imposed by the need to preserve incentives to efficient supply and necessary expansion." In letters and conversations throughout 1962, in particular, Cowperthwaite pushed Kadoorie to accept a merger or, at least, to equalize electricity rates across the colony. He also proposed that the government take shares in the merged electricity supply company.[102]

Cowperthwaite appears never to have articulated why he believed that consumers should benefit from the money they paid CLP, except to intimate that CLP was abusing its monopoly position and that the government needed to protect consumers. It is a curious view for someone who espoused the idea of positive noninterventionism, the notion that the government should not intervene unless a strong case could be made for action. Cowperthwaite's view was simply asserted with no attempt to justify it. This omission is telling and suggests that the standards for acceptable corporate behavior were much stricter in the immediate postwar period than in the post-1980 decades. CLP director Sydney Gordon told Cowperthwaite that it was nonsense to propose that all reductions in retained earnings should be given to consumers.

Gordon warned that expansion of the electricity supply system could be halted as a result. Cowperthwaite responded that this was "impossible.... Government wouldn't allow [it]." He added that the suggestion was "absurd.... It illustrates what can happen when a private Company has a virtual monopoly."[103] On another occasion he discussed the dangers of monopoly profits:

> One of the basic findings of the Electricity Supply Companies Commission was that it is improper that a private company engaged in public utility operations on a *de facto* monopoly basis should finance the expansion of its undertaking from profits taken from consumers in excess of a reasonable return on capital subscribed by shareholders and then distribute additional dividends out of the additional profits accruing from this expansion.[104]

Cowperthwaite's notion that profits ("reserves") did not belong to shareholders was a radical challenge to CLP's financial business model. And he was unyielding in his insistence that consumers should not pay for new capital investment:

> Mr. Cowperthwaite considers that reserves do not belong to the shareholders and should not be available for the issues of bonus shares or for dividend purpose. It was pointed out to him that the issue of bonus shares made no real difference provided the dividend was adjusted. Also there was no doubt that the reserves in the Company's accounts belonged to the shareholders. In the event of a winding-up who else would have a claim on them? Mr. Cowperthwaite said that he might agree to this view as regards reserves "prior to about the middle of 1960" but he could not agree as regards subsequent reserves which had been built up due to Government having frozen dividends and reductions in rates. He felt very strongly ... that if at some future

date Government compulsorily acquired the Companies they should not have to pay for reserves or assets built up out of these retentions.... Mr. Cowperthwaite indicated that if we did not give way on this paragraph he would recommend nationalisation of the Companies.[105]

Cowperthwaite may or may not have wanted nationalization. But he discussed the specifics of nationalization and the amount of compensation with the companies at a level of detail that made the threat of it real enough to win the concessions on finance and governance that he wanted.[106] Kadoorie repeatedly claimed to his associates that the ESCC had embarrassed the government by going too far in recommending nationalization: "The reason the Companies have made an 'inspired' informal approach to Government is that the findings of the commission have caused embarrassment to both Government and industry who well know that the whole future of The Colony lies in the word 'freedom,' a warning I conveyed to the Commission in the closing words of my address: 'No one should treat the responsibilities of this heritage lightly, lest in destroying these freedoms they destroy the Colony itself.'"[107] Privately, he was less sure. After meeting with Kadoorie in London in June 1960, one of his advisers wrote that "the Companies still envisage the possibility of nationalization if negotiations on the scheme break down."[108]

Cowperthwaite, as noted, took a tough line in negotiations. "Mr. Cowperthwaite considers that reserves do not belong to the shareholders ... [and] indicated that if we did not give way on this paragraph he would recommend nationalisation of the Companies"[109] As late as February 1962, Lawrence confided to Horace in their private diary that he feared CLP would in the end be nationalized.[110]

In the United States, Insull had sold bonds to finance Chicago Edison in London as early as 1896.[111] In Hong Kong, this option did not exist. In the period immediately after the war CLP ran a small overdraft with HSBC, and HSBC had guaranteed payment on the generator that was ordered immediately after the end of hostilities. CLP relied almost exclusively on cash flow and shareholder investments to finance expansion and was effectively debt-free, an extremely unusual accomplishment for an electricity supply company.[112]

On June 7, 1962, CLP shareholders met to consider the proposed merger with HKE.[113] The board, under pressure from the government, had recommended the merger to shareholders. As chairman, Kadoorie had to support the merger on the basis of a share exchange ratio that did not place much value on CLP's growth prospects.

It is clear from an examination of the archive that Kadoorie was working to help opponents of the transaction he had agreed on with Cowperthwaite. Not a single shareholder spoke in favor of the proposed merger. As noted, proxies opposing the merger, combined with an informal poll at the meeting, allowed Kadoorie to state that it would be impossible to get more than the required supermajority of 90 percent of shareholders needed for the merger to be completed. Kadoorie thus adjourned the meeting without calling a vote. This ensured that the Kadoorie family did not need to vote its 39 percent stake.[114] A vote against the transaction by the Kadoorie family's shares would have stopped the transaction but could have provoked retaliation by Cowperthwaite.

Cowperthwaite was nonetheless annoyed at the result. He had no sympathy for short-term shareholders, whom he accused of being speculators because they had not invested capital in the company itself. They had not lost "real money," he wrote Kadoorie.[115] In a subsequent meeting with the finance secretary, Kadoorie said that he "had thought shareholders might be indifferent, in which case it was wise to make the best of a bad bargain."[116]

As we have seen, Cowperthwaite was insistent that CLP end its practice of financing capital expansion using retained earnings and dividend payments. He regarded this money as rightfully belonging to consumers. He was adamant that the company should instead go to the public share market to raise funds. Cowperthwaite had a static view of capital. He believed that there was only a finite amount in the economy at any given point; he wanted to free up capital by having CLP, one of Hong Kong's only internationally creditworthy companies, borrow money from the World Bank. CLP was primarily raising funds from internally generated cash from operations and only occasionally from existing shareholders. Why Cowperthwaite believed that giving electricity rebates to consumers and raising more money from the World Bank would solve Hong Kong's shortage of capital deserves more research. He appears to have implicitly believed that rate cuts and CLP dividend payments would provide capital that would be invested elsewhere in Hong Kong.

ELECTRICITY AS A POLITICAL PROJECT

There are other equally puzzling issues that center on Cowperthwaite's conception of the role accorded to capital and the rights of owners of capital. He had no sympathy for investors who paid more than the par value of shares, contending that any money beyond par value was simply speculative capital, as it did not go to the company to be used for productive investment. Though supposedly an apostle of free-market capitalism, he had a cavalier attitude about shareholders' rights, believing that the company's owners had only the most minimal claim to money that electricity consumers paid to the company. The fact that $250 million was wiped off the market capitalization of CLP during the debate over nationalization was of no concern to him. He seemingly never acknowledged the fact that Robert Shewan, the Kadoories, and others had built the company over six decades. He consistently advocated nationalization through the stock market, answering Kadoorie's fears that the market would not be able to absorb such large issues by promising that the government would buy any unsold shares and that it would commit to being a long-term shareholder of the company. His actions are at odds with Milton Friedman's hagiographic portrait of Cowperthwaite; they reflected his pragmatic approach as financial secretary to finding capital for continued economic development.

Chapter Six

"DIE-HARD REACTIONARY" IN THE EXPANDING COLONIAL STATE

1964–1973

China Light & Power's existence as an autonomous company controlled by Lawrence Kadoorie came under sustained attack for five years after the 1959 ESCC hearings. The Hong Kong government, with Financial Secretary John Cowperthwaite in the leading role, determined that new controls would be imposed on the two electricity supply companies, with more of the burden falling on the faster-growing CLP. The only question was what form those controls would take. Nationalization, which would include a compulsory buyout of shareholders, remained a possibility for which CLP's board of directors prepared. Whether or not nationalization occurred, merger with rival HKE appeared almost certain. Hong Kong's lack of government oversight of electricity supply companies had long made it a global outlier. No longer. Government control and reduced electricity tariff rates were quickly accepted as inevitable after the release of the ESCC report.

CLP's large profits, its confusing surcharges and tariffs, and its need for capital to expand came at a time when the very idea of privately owned electricity supply companies was under attack globally. Even in Hong Kong, a colony that prided itself on minimal government interference in business, public opinion and policy shifted against the two monopoly suppliers of electricity now that the electricity CLP and HKE produced was no longer seen as a luxury but a vital public good. By this time the idea that electricity was a natural monopoly—a business where an established company had

"DIE-HARD REACTIONARY" IN THE COLONIAL STATE

barriers to entry that made competition virtually impossible—was well accepted in Britain and the United States and underpinned general acceptance of regulation both of price and terms of service. This belief in the necessity of regulation developed in Hong Kong as part of a general sense that electricity was too expensive and too important, and the electricity supply companies were too big to be left unregulated. The prevailing thinking among senior government officials, editorial writers, civic leaders, and the heads of some Chinese business associations was that electricity needed to be subject to government control. Some sort of permanent commission to provide oversight of the electricity supply companies seemed likely.

In the aftermath of the ESCC hearings and the commission's recommendation for nationalization, Lawrence Kadoorie and other directors grudgingly accepted the need for some form of control as a way of securing access to finance for expansion. The government's continued restrictions on dividend payments were the stick it wielded to enforce change. The government did not use legal means to compel CLP to freeze dividends; the mere threat of action against the directors had the required effect. The dividend limit meant it was impossible to pay out profits. CLP could not continue as it had before 1960 if it wanted unimpeded access to finance. As noted, the hearings and subsequent negotiations took substantial amounts of Kadoorie's time for five years.

Kadoorie found a solution to the government pressure by bringing in Esso as a joint venture partner. As part of the agreement, CLP accepted restrictions on profitability, cut its electricity tariffs, and made electricity charges more transparent. The cap on dividend payments was lifted. Capital that had accumulated during the dividend freeze was freed up for new investment. Esso, formally known as Standard Oil of New Jersey, became the world's largest oil company when John D. Rockefeller's Standard Oil Trust was broken up by the U.S. Supreme Court in 1911 (it comprised 43 percent of the trust's value).[1]

In 1964, Esso remained one of the world's largest oil companies, part of the Seven Sisters group of companies that dominated global petroleum production at the time. Before it was broken up, Standard Oil had set up in Hong Kong in the early part of the century, building facilities in Lai Chi Kok.[2] In the 1960s these were operated by Mobil, another company in the former Standard Oil group. Coincidentally, as Kadoorie was looking for capital, Esso's joint venture marketing arrangement with Mobil in East Asia was ending. Esso, the weak sister in the region, was in search of significant populations of long-term customers.

The rationalization of electricity tariffs and the acceptance of government oversight was a significant departure from the pre-1959 period and an even more radical move away from the pre–World War II electricity order. Electricity in Hong Kong and Guangdong had until the 1930s been attractive for speculative investors and was often an element in a portfolio of mostly unrelated businesses, as seen in the case of Shewan Tomes. The need for continuing capital investments in the electricity supply business prompted the Kadoories' takeover of CLP's management in the early 1930s and the subsequent expansion at Hok Un that was completed in 1940. The Kadoories' ascendancy led to management specialization and ushered in a more technical and capital-intensive phase of the company's history.

The 1945–1964 period was one of rapid expansion. These two decades exposed what proved to be CLP's unsustainable business structure, that of a company that was subject to no social or political control at a time when electricity demand and concomitant stresses were growing quickly. At the same time, the technology of electricity supply, the fossil-fuel resources used for generation, and access to adequate generating and distribution capacity were all in flux. The lack of stability was exacerbated by the larger social and economic changes previously discussed. The ESCC hearings and recommendations for nationalization and merger and their aftermath marked a period of extreme instability and posed the very real danger that the Kadoories would lose ownership and management control of CLP.

As the views of government officials, journalists and editorial writers, civic organization leaders, and even other business executives shifted against the company, Kadoorie modified his thinking. Rather than see the company nationalized or merged, Kadoorie moved to a middle ground and accepted the need for control of tariffs and profits. Cornered by a government which refused to sanction dividend increases, Kadoorie reached halfway around the world for a fresh source of capital. The decision to bring Esso in as an investment partner represented a notable example of a dynamic response to resolve a shifting, unstable, and threatening situation.

A tension was evident between the technological modernity that CLP embodied and promoted and the countervailing political and social challenges that this very technology created. CLP as a company was slow to appreciate the social and political changes that its electricity made possible. How did Kadoorie manage to reconcile government and public demands for more control of CLP with the need for additional capital and an expanded electricity supply system? How did the promise of new capital and technology mesh with the reality of a city crowded with refugees living in squatter settlements? The answers to these questions show how CLP, with financial and technical support from Esso, and regulatory and administrative assistance from the colonial administration, inserted and enmeshed itself more fully in Hong Kong households and businesses during the period under review in this chapter. The process changed CLP, making it a larger and more technologically, financially, and managerially sophisticated company. CLP in turn continued to change Hong Kong, providing the ever-increasing amounts of electric current needed to sustain Hong Kong's economic expansion.

ESSO AND THE SCHEME OF CONTROL AGREEMENT (SoCA)

CLP, Esso, and the Hong Kong government announced on November 23, 1964, that the U.S. oil giant would invest $220 million in a venture with CLP to build a 720-megawatt oil-fired electricity-generating facility. This was the largest foreign investment ever made in Hong Kong and was seen as a vote of confidence by a U.S. multinational. Esso gained a significant new customer for its oil and an agreement with the colonial government that would allow for a profitable return on its capital investment. Esso's vice president, Charles Leet, said in the announcement that the company was "very much impressed by the free enterprise climate maintained by the Colony's Government and by its energetic and progressive business community."[3]

In exchange for an easing of government restrictions on dividends, CLP agreed to cut electricity tariffs by an estimated $18 million the following year and to limit profits to 13.5 percent of invested capital. Any excess money would go into a development fund to invest in expansion. For its part, the government said that it wanted to limit profits to a "reasonable return" and to make sure that consumers benefited from any additional expansion in the form of lower electricity rates. This was accomplished with an explicit division of future revenues from electricity sales between consumers (who would benefit from lower tariffs, since their rates were reduced because of a restriction on the profits that CLP could earn), shareholders (who would enjoy higher dividend payments as a result of growth), and a capital investment fund to pay for new electricity generation and transmission facilities. No longer would Kadoorie and the board decide what rates they would charge for electricity.

The new 720-MW electricity-generating facility Esso and CLP announced was originally planned to be about twice the size of CLP's existing generating capacity. In fact, higher than expected growth required an even larger unit. When the last phase of the plant was completed a decade later, total generating capacity at the Tsing Yi facility was 1520 MW, more than two times larger than originally planned.

In addition to making CLP a much larger company, the agreement changed it in other ways. The SoCA also resolved the five-year standoff with the government over control of the colony's electricity supply business. In return for limits on profits, the shareholders retained control of CLP. As chairman and steward of almost 40 percent in shareholding, Kadoorie's stake was by far the largest and most influential. This newfound stability, which provided the security of a government-sanctioned profit arrangement, allowed the scale of CLP's operations to expand tenfold in the next two decades. Moreover, Esso's management expertise infused CLP with new business techniques and controls that proved important as the company grew rapidly. Michael Kadoorie said that CLP "deliberately and specifically used a lot of Esso techniques as a way of making a more sophisticated company."[4]

The signing of the SoCA and the agreement with Esso showed Kadoorie at his most strategic; he drew on a global network of contacts that provided capital, technology, and expertise to strike an arrangement that constituted one of the most important moments in CLP's history. The Esso relationship not only infused CLP with capital, technical, and managerial expertise but brought CLP and Hong Kong more fully to the attention of the U.S. government.

CLP switched from burning coal to burning oil at its electricity-generating stations in the late 1940s. With electricity demand growing almost 20 percent annually during much of the 1950s and early 1960s, the company was a large and growing customer for oil. Lawrence Kadoorie reasoned that one of the big international oil companies might be willing to invest in a joint venture to build an oil-fired electricity-generating plant in fast-growing Hong Kong as a way of securing a long-term customer.[5]

He initially approached Mobil, which already had a refining facility in Hong Kong, but the company turned him down. Next he met with Esso, which had no investment in the colony.

Negotiations started in July 1962 when Kadoorie met Esso executive George Bell, who had earlier been based in Hong Kong and was on a round-the-world tour to try to sell oil. The memo Kadoorie wrote on July 4, 1962, was aptly titled " 'Desperation Point' Generating Station & Refinery" and labeled "Strictly Private & Confidential." Desperation Point seems to have been a metaphor (there is no location with that name in Hong Kong), one that aptly summed up where the company stood.

Kadoorie outlined an agreement according to which Esso would put up two-thirds of the capital for a new industrial facility comprising an electricity-generating plant, a petroleum refinery, and perhaps salt-water distillation capabilities. Under Kadoorie's proposal, CLP would invest the remaining one-third of the funds. Kadoorie personally signed the memo with a green-ink fountain pen, a flourish he generally reserved for important documents. At or immediately after this initial meeting, Kadoorie sketched out on a yellow legal pad the outline for an agreement. The original memo and sketch were remarkably similar in general outline to the formal document signed sixteen months later.[6] Two weeks later, on July 18, Esso President J. W. Pickering wrote to express his "considerable interest" in Kadoorie's proposal.[7]

Kadoorie knew from the start that an Esso investment would have global geopolitical ramifications. He had long been weaving ties between Hong Kong and the wider world, especially Britain. Now, for the first time, he brought in the United States. "It will be appreciated that a project of this nature requires negotiation at the highest level and involves international commitments," he wrote in his initial Desperation Point memo. "When considering the broader aspects," he added, "participation by Government and the World Bank with leading American interests should be anticipated. Provision of Hong Kong's basic power need would no longer be a domestic matter."[8]

Kadoorie's implication was that U.S. investment would provide a form of protection for CLP, since the colonial government would be loath to interfere with the management of a company in which a major U.S. oil company was involved. Writing in the *Financial Times*, Kadoorie maintained that Esso's investment was a vote of confidence in Hong Kong's political economy. "International cooperation and a commitment of American equity on this scale is a welcome proof of confidence in the Colony's future prosperity and represents an important contribution to Hong Kong's rapidly growing needs for outside capital investment."[9]

Just as Kadoorie saw the investment in global terms, so, too, did the U.S. government. As noted, Eisenhower had recently agreed to use U.S. forces to defend Hong Kong in the event of an attack by the PRC.[10] Around this time, the U.S. government's attitude changed from anger at Hong Kong's part in undercutting sanctions imposed against China as a result of its involvement in the Korean War, to, in the words of historian Priscilla Roberts, appreciation of "the international propaganda value of Hong Kong as a flourishing capitalist redoubt, an oasis of economic prosperity and development that offered ever-greater contrasts and an implicit rebuke to the bleak situation on the Communist mainland."[11] The electricity the new Esso-CLP venture provided would make its streets and shops brighter and provide even more of a contrast with the dark mainland.

The Esso investment in CLP, by deepening U.S. involvement in the colony, fit into this strategy of bolstering Hong Kong as a shining city on the harbor surrounded by a Communist hinterland that lived in darkness and shadow. The presence of one of its largest and most influential corporations gave the United States a tangible stake in the future of Hong Kong. Esso's investment in the new electricity-generating plant would aid the colony's continued rapid expansion, making possible continued growth in one of the world's fastest-expanding economies. It would be a propaganda coup for the capitalist world and, by providing the electricity for factories that provided jobs, foster social stability. Hong Kong's workers would also consume electricity, both at home and in shops and restaurants.

The significant fixed investment represented by an electricity-generating station would make it impossible for an American company to simply close down during a crisis, as Chase bank had done in the initial stages of the Korean War. Kadoorie nonetheless repeatedly expressed concern to Esso executives during the negotiations that the United States might use its sanctions power to cut off the supply of oil to CLP's new electricity plant.[12]

The geopolitical sensitivity of the investment was reflected by Esso's request, before finalizing the agreement, for a ruling from the U.S. government that it would not be penalized for breaking the trade embargo with China by selling electricity to Communist Chinese nationals. On

January 10, 1964, in a letter headlined "Request for Ruling Under Foreign Assets Control Regulations," Leet wrote to the U.S. Treasury Department that Esso was considering investing approximately US$90 million in a joint venture for power generating, fuel terminaling, and storage facilities but warned that it could not control who bought the company's electricity:

> It is believed that such a project will represent the first sizable United States Industrial investment (outside of petroleum terminals) in Hong Kong.... Esso has no means of investigating or controlling the choice of China Light and Power company's customers. It is possible that there are now or may be in the future persons or organizations among China Light and Power Company's customers in Hong Kong who fall within the prohibited categories under the Foreign Assets and Foreign Transactions Control Regulations.[13]

Communists, in short, could be among those using the new company's electric current. Kadoorie did not appear to care and quipped that half his customers were probably Communists. He ridiculed the Americans for being "scared to death of the word Communist," noting that CLP was twelve miles from the border of the world's largest Communist country "and if I could sell electricity to them, I would."[14] Kadoorie knew he was viewed by some people as a "die-hard reactionary," but he did not care what his customers' political beliefs were.[15]

Neither, it appeared, did the U.S. government. The joint venture would only own the generating plant, for CLP had agreed to buy and distribute the electricity produced using Esso's oil. CLP was plainly not a Communist Chinese entity. That was enough of a cover for Washington. Just three weeks later, on January 29, U.S. Treasury Director Margaret Schwartz wrote to Leet saying that the government had no objection to the arrangement provided that no "equipment, supplies, services, etc. of any nature whatsoever" were obtained by the proposed joint venture from a "national of Communist China or North Korea." No license would be needed for the new company to sell electricity to CLP for distribution in Hong Kong, even though Communist Chinese nationals could be among CLP's customers.[16]

Treasury could have decided that, just as Communist China nationals could not supply equipment to the venture, they could not buy its electricity output. Its quick decision to allow the venture to go forward suggests that the Esso-CLP joint venture, known as the Peninsula Electric Power Co.

(hereafter PEPCO), fit into the pattern described by Priscilla Roberts wherein Hong Kong's prosperity was something to be nurtured by the United States government given that Hong Kong's ultimate loyalty to the West was no longer in doubt. PEPCO's plant, built on the island of Tsing Yi, was successful far beyond Kadoorie's and Esso's original ambitions. The first unit opened in 1969. By 1972, the Tsing Yi facility was producing about half of the electricity in the CLP system. During the middle and late 1970s the Tsing Yi station produced almost all of CLP's electricity.

This was a project that allowed for the continued expansion and transformation of the Hong Kong economy. The investment also changed the narrative about how an electricity supply company should be controlled. In the late 1990s, Stephen Goldman, chairman of the successor company, Castle Peak Power Co. and chairman of Exxon Energy Ltd., declared that the agreement "halted discussions, prevalent in Hong Kong at that time, on the possibility of nationalization of the power companies in Hong Kong."[17] The Esso investment, in ending talk of nationalization, also brought to a close the era of electricity as a political project. Henceforth electricity would primarily be a financial and technical endeavor.

ELECTRIC POLITICS

Kadoorie saw the dramatic changes taking place in Hong Kong in a historical, political, and economic context. Writing in the *Financial Times* in 1965, as construction for the Tsing Yi project was just beginning, he lauded Hong Kong's "relatively painless" electric-powered "industrial revolution" and its ability to forestall "unemployment and civil unrest" that would have threatened the colony's very existence:

> Political events in China, the closure of the American market to Chinese goods, and the simultaneous arrival of refugees with technical knowledge and, in some cases, with capital, have all contributed to an industrial revolution which has been relatively painless. This has been mainly because the motive power for production, the different means of converting primary products into finished materials, were available when and where required. The importance of electricity in this context cannot be exaggerated. Without an adequate supply industry would have been hampered, and unemployment and civil unrest would have plagued the Colony's security and undermined

its existence.... The Government's enlightened policy of free trade and of encouraging private enterprise has been amply justified in all fields, and the colony's unparalleled rate of industrial expansion has only been possible because of the power companies' intelligent anticipation of the enormous increase in the need for electricity and their ability to keep ahead of demand.[18]

The article shows how Kadoorie situates the Colony and, implicitly, CLP, on a larger historical canvas. It also demonstrates how he promotes himself as a revolutionary at the forefront of unparalleled historical change.

Kadoorie's claim that Hong Kong was in the midst of an industrial revolution equates Hong Kong's industrialization in the 1960s with the Industrial Revolution launched in eighteenth-century Britain. By the 1960s, Hong Kong's textile industry was a significant exporter to the United Kingdom, and expansion was underway in other export-oriented industries, such as apparel, electronics, and plastics. But in claiming that this was an industrial revolution, Kadoorie ignores the fact that there were no significant advances in technology or production processes. He also ignores the longer history of Hong Kong's servicing of the empire. Hong Kong benefited from imperial trade agreements, notably the imperial preference system that was agreed upon at the British Empire Economic Conference in Ottawa in 1932.

Hong Kong's high economic growth was a process of industrial catch-up, not of revolution. Kadoorie situates electricity at the center of this historical process, but in doing so he conflates the original industrial revolution, which took place before the advent of commercially produced electricity and relied on water and steam for motive power, with the very twentieth century transformation that depended on reliable supplies of commercially produced and widely distributed electricity. He also fails to note that, as early as 1940, colonial authorities had counted some 800 Hong Kong factories.[19] After the war, Hong Kong exports in the 1950s and 1960s relied on cheap labor, increased capital intensity, and the physical dislocation caused by the Chinese Communist victory in 1949, which brought Shanghai textile businessmen to Hong Kong.

Shifting to a broader political and social frame, Kadoorie contrasted Hong Kong's beneficent "industrial revolution" with the threat of political revolution. Jobs and incomes were needed to prevent an uprising, especially from the more than one million refugees who had come to the colony and had little to lose from continued upheaval. Kadoorie maintained that a free and open economy could best provide those jobs and in so doing neutralize social unrest. Here he was implicitly rebuking the arguments in favor of nationalization. Adequate supplies of electricity underpinned economic growth and job creation, ultimately reinforcing social order. "Unemployment and civil unrest would have plagued the Colony's security and undermined its existence," he claimed, without wise policies in the form of "free trade and encouraging private enterprise" as well as "the power companies' intelligent anticipation of the enormous increase in the need for electricity."

Kadoorie here consciously presents himself as the heroic entrepreneur and visionary who is not just providing a commercial service but is, with the help of the "Government's enlightened policy of free trade and of encouraging private enterprise," following in a tradition that, as his choice of words implicitly reminds readers, can be traced back to the Enlightenment. Kadoorie was styling himself as the heir to a tradition that has at its very center the notion of light—light that his electricity supply system provided. Electricity for Kadoorie was more than electrons on the move; the free flow of electricity facilitated the circulation of capital, people, and technical expertise. For Kadoorie, Hong Kong was a place where flows of electricity contributed to flows of capital and people, with increases in one fueling growth in the other.

CLP, as a large, capital-intensive company, faced strikes and consumer boycotts throughout its history. CLP's Canton operations were the targets of a boycott organized to protest the permanent extension of the U.S. Chinese Exclusion Act in 1904. As late as 1977, Kadoorie would marvel at the unfairness of this action, despite evidence that, from the boycotters' perspective, the company was an understandable target.[20] CLP was swept up in larger political and social movements, such as a lengthy strike in 1925 that affected businesses throughout the colony; the company paid staff bonuses in compensation.[21] In the late 1940s, it was the subject of a fifty-day strike that was settled with the negotiating help of board member M. K. Lo and the female leader of a group of strikebreakers. "They were able to get a woman, a very tough character who claimed to be a pirate from Bias Bay, with a gang of men under her," remembered Eugene Joffe in a 1979 interview. "The strike was settled pretty much on CLP's terms. . . . There were no more strikes until '67."[22]

"DIE-HARD REACTIONARY" IN THE COLONIAL STATE

Mao's Cultural Revolution spread into the colony in 1967. There were bombings, some fatal. Large-character Chinese posters attacking the colonial government were hung out of the upper-floor windows of the Bank of China building in Central. Coincidentally, Kadoorie's headquarters, the St. George's Building in Central, was being rebuilt. Kadoorie ordered construction crews to work overnight, illuminating Central in a blaze of defiant light in order to "inspire confidence in the future of Hong Kong."[23] During the construction of the Tsing Yi power station, probably in 1967, when a contractor was changed, "a strike erupted and all sorts of threats were made." Staff were held hostage and prevented from leaving the site until the matter was settled.

Alex Buchanan, Tsing Yi's first station superintendent, remarked that they were fortunate to have been able to resolve the situation "comparatively quickly" given the "political climate and the militant labor unrest prevailing at that time." Buchanan went straight to CLP headquarters after the siege at Tsing Yi, sporting four days' beard growth. "The next day I received an electric razor with a note saying 'Whenever you go to Tsing Yi, take this with you.'"[24] The electric razor likely came from Kadoorie, a gesture characteristic of his sense of humor; moreover, no one else at CLP would seem to have had either the wit or the resources to buy and immediately dispatch such an item.

The company's operations suffered during the 1967 upheaval, though the archival record is thin. On the first night of the 1967 riots, a turbine shed a blade: "Almost no labor being available, a team of engineers buckled to and started repairs."[25] As late as May 1968 there were still "quite a few labor problems," Wood noted in a letter.[26]

Kadoorie by this time was more comfortable asking for government help. In early 1968 he dismissed the fading leftist campaign of violence as "nuisance bombings," even as he unsuccessfully sought insurance against terrorism.[27] When he could not obtain the insurance from the private sector, he turned to the government for help. Financial Secretary John Cowperthwaite, in a letter of January 16, 1968, offered Kadoorie some hope that help would be forthcoming if needed. Cowperthwaite wrote that although the government would not guarantee to indemnify CLP against a terrorist act, the company could "make representations to Government and these would be considered in the normal way and in light of all the circumstances then prevailing. At the same time I know you are aware that your company's installations are regarded as quite vital to the normal life and security of the Colony and, should the need arise, Government will do everything within its power to protect such installations against damage."[28]

Although most workers were Chinese, a small cadre of expatriate managers formed an important part of CLP's makeup. In 1955, the year for which we have the most complete data, twenty-eight expatriate managers oversaw a workforce that included fifty-five "local" employees (most of whom were ethnically Portuguese) and 703 Chinese. Summaries of the work careers of eighteen mostly expatriate staff which were prepared for an ESCC visit to CLP facilities show diverse backgrounds. Most had worked elsewhere in the British Empire, or at least outside of Britain, from Southeast Asia to Brazil. They had been employed by a variety of companies ranging from other electric utilities to American Express. Although many had technical backgrounds, the ability to thrive as an expatriate manager was also important. One executive recruiter ranked prospective employees on, among other factors, their ability to get along with local servants.[29]

Station manager Cyril Wood had trained in mechanical and electrical engineering with the Doncaster Colleries Association before working at Callenders Cable and Construction and switchgear supplier A. Reyrolle. He joined CLP as a district engineer in 1930. Accountant Alan Onslow had worked for the Brazilian Traction, Light & Power Co. in Rio de Janeiro in 1950–1951 and as a traveling auditor for American Express in India, Pakistan, the "Far East," and Australia from 1952 to 1955 before joining CLP in 1957. Distribution Engineer J. W. Barker had worked as an assistant distribution engineer in York, England, before joining CLP in 1939; during the war, he was interned in Hong Kong and Japan. Acting Distribution Engineer C. S. Rolfe had worked with the British Electricity Board before joining CLP as a maintenance engineer in 1948. During the 1950s Meter Superintendent G. H. V. Ribeiro, who had joined the company in 1932 after a five-year apprenticeship, and Technical Assistant Kao Sun Zee were sent to Britain for further training.

ELECTRICITY AND TECHNOLOGY IN THE COLONY

Many new technologies spread quickly to Hong Kong. Most of these originated in or were transmitted through Britain. As noted, commercial electricity generation in Hong Kong began eight years after Edison's successful

"DIE-HARD REACTIONARY" IN THE COLONIAL STATE

demonstration of the feasibility of transmission and distribution in London and New York City. Other technologies typical of modern urban life appeared in Hong Kong fairly quickly. Hong Kong's connection to leading global centers by steamship, telegraph, and, beginning in the 1930s, airplane, facilitated technological diffusion. Hong Kong, like Bombay, Calcutta, Melbourne, and Singapore, displayed technological sophistication. It should also be noted that this sort of leading-edge technology in Hong Kong was limited to a relatively small group of people and concentrated in the non-Chinese areas of the colony.

New technologies met with little apparent resistance. The introduction of electricity faced less opposition in urban Hong Kong than it did in Texas, where farmers were terrified of it. Although the advent of electricity and other novel technologies was noted by newspapers, the reaction was generally matter-of-fact acceptance.

New technologies that were celebrated constituted the exception. The *SCMP* in August 1957 used a front-page story to showcase the introduction of the colony's first escalator in a public building. The escalator, in Central's Man Yee Building, was presented as if it were an amusement park attraction:

> Business was brisk, with queues waiting to ride the escalator—mainly for pleasure. Old Chinese women and little pig-tailed girls with babies on their backs were amongst those to try the novelty. Four police constables, two at the top and two below, stood by to keep order and see that no one had too many rides. The expressions on some of the faces ranged from sheer delight to apprehension. Others rode with a look of superior indifference. The first step caused a few to wobble but Otis man [*sic*] were there to give a steadying hand and words of advice.[30]

The newspaper contrasted the new world of quantified technology—readers were informed that the escalator rose nineteen feet six inches vertically and was designed to carry five thousand people an hour at a distance of ninety feet a minute on the incline—with the traditional world of "old Chinese women" and "pig-tailed girls with babies on their backs." The Otis man or men and police constables were there to offer support but also to ensure that no one had too much fun. Otis Elevator manager Colin J. Ure, whose company installed the escalator, "said there was a big future for escalators in the colony." The colony's first six escalators, also manufactured by Otis, had already been installed at the Jockey Club. Twenty more were being considered for other buildings.[31]

Ascribing the sort of magical, universal powers that Robert Fraser-Smith did to electricity was the exception. The electricity produced and distributed by CLP and HKE embodied and enabled technological modernity. CLP and HKE were among the largest and most sophisticated users of technology in Hong Kong. They also made possible products and services as diverse as ice, air conditioning, lighting, water pumping, elevators, and factory machinery for their customers.

Electricity generation technology was viewed by Kadoorie and his peers as something that was constantly changing and improving. The electricity supply business was one in which ever-larger investments were needed to take full advantage of technological progress. In the 1930s, CLP changed the generating frequency from 60 to 50 cycles and the voltage of low tension distribution from 110 to 200 volts for lighting and from 200 to 346 volts for power and heating. These were significant technical transformations.[32] The switch from coal- to oil-fired generating units, and the possibility that nuclear-generated electricity would make existing fossil fuel-fired plants obsolete, generated a sense of technological flux and instability.[33] Innovation continued at a rapid pace. Hong Kong's electricity supply companies watched and often imitated global changes. In his 1957 letter to shareholders, Kadoorie wrote: "We must remember that in no industry is there a greater premium on finding new and better ways of doing things. This constant search involves every major field of science and technology, and its accomplishments are felt throughout [the] world economy."[34] "New and better ways of doing things" summed up Hong Kong's technology-led economic, political, and social program. CLP and HKE were technological pacesetters. They introduced new technologies that later became more widely diffused in the colony, much as funding from the military in the United States led to the use of technologies that later were used widely by the general populace. Communications technologies were important to managing electricity distribution networks. Beginning in 1950, CLP's workers used VHF radios to communicate between remote parts of the colony and the company's headquarters. The 1953 annual report cited the radio's role in minimizing time lost during repair work after Typhoon Susan. Telephone networks were constructed. In 1960, CLP noted that the company had "our own direct telephone lines between the Administration Building and the Power Station, as well as to all key Sub-stations."[35] At the time, there were many complaints in newspapers that the Hong Kong telephone system was inadequate. CLP contrasted the record of a monopoly with its own successes in building a private telephone network.[36]

In 1946 the company's meter readers still used abacuses.[37] New calculating technologies were needed in order to accurately bill and receive payments from a growing number of customers. Manual calculators and then electronic calculators followed. In 1953 the company ordered a "Public Utility Billing Machine" to meet its continued expansion needs.[38] In 1963, CLP introduced the colony's first commercial computer.[39] Less obviously, new technologies included those used in electricity generation and distribution. These were promoted in the CLP and HKE annual reports. In 1960, CLP doubled the voltage from 33,000 to 66,000 volts in its main feeder lines.[40]

CLP worked to foster technical skills in Hong Kong. The company donated $200,000 in 1957 to the Hong Kong Technical College, the precursor to the Hong Kong Polytechnic University, when it moved from Wanchai to Hung Hom, close to CLP's electricity-generating station. In making the donation to build and partially equip what would become the China Light and Power Co. Electrical Machines Laboratory, CLP cited its need for skilled technicians. It declared that the laboratory "will help industry by training youth for operating and maintaining machines."[41] The Chinese Manufacturers' Association and the colonial government each donated $1 million to finance a substantial expansion of the college. Economist Edward Szczepanik described it as "the principal government institution for training in mechanical engineering, building, telecommunications, navigation, and commerce."[42]

In 1960, CLP took concrete steps to improve its in-house knowhow by allying with British industry groups. The company joined the British Electrical and Allied Industries Research Association and noted that it had "received much useful technical information from this source." Kadoorie also noted that CLP had been in close touch with the British Electricity Council.[43] These contacts broadened CLP's executives' knowledge of the technical and regulatory aspects of the industry and reduced their reliance on expertise from suppliers, architects, and consultants. The Esso joint venture accelerated this learning process.

One significant difficulty in the largely successful importation and implementation of technology was the tyranny imposed by distance. This was a recurring challenge from the earliest years of electricity in Hong Kong and appears again as an issue during CLP's expansion of Hok Un in the late 1930s, the earliest period for which extensive internal corporate material is available. In one of many complaints about the difficulty of ensuring timely deliveries of equipment, shareholders at the 1948 annual meeting were told about the "ever-prevalent difficulty in obtaining materials from England, and in particular the steam piping which did not arrive until mid-October, [meaning that the turbine] installation was seriously delayed."[44]

References to problems in the HKHP archives and CLP annual reports abound during the 1950s. There are repeated mentions of the difficulty of getting timely delivery of replacement parts and of the generating equipment itself, and of needed repairs. In 1953, Kadoorie reported that the No. 10 unit had "manufacturing problems." He complained of "unsatisfactory progress" and noted that the "Erection Engineer for this Turbine will shortly be arriving in the colony from the United Kingdom."[45] Problems of distance were less apparent during the construction of the Tsing Yi project, despite the much larger size of this new facility, than in the Hok Un expansions of the 1950s. This could reflect the greater organizational and managerial capabilities of Esso. Jet travel, which allowed for faster, less expensive movement of people and goods (particularly after the introduction of Boeing's 747 in 1969), also played a role in effectively collapsing the disadvantages of distance.

The introduction of new technologies was likely even more problematic than that suggested by the contemporaneous records to which we have access, especially those at the HKHP and other archives. A 1966 article in *Asian Industry*, three years after CLP's NCR computer was installed, extolled the "new system for controlling stores and wages and for greater speed and efficiency in preparing bills and accounts. The computer also works out the trends in industry."[46] Only in an oral history done some fifty years later do we find that the introduction of the NCR computer in 1963 involved a "very difficult" transition. "It didn't work very well" at first, remembers Patrick Bailey, who worked alongside chief accountant Alan Onslow and recounted the difficulties in switching from a mechanical to a computerized system.[47] Both of these statements could be true, for by 1966 CLP staff might have made effective use of the computer to prepare bills and accounts, and perhaps to work out industry trends as well. Still, Bailey's retrospective comments are a reminder that archives tend to smooth off many of history's rough edges.

The use of electricity had been an expensive luxury before World War II. In the 1940–1941 Administration Reports, the governor's budget for fifteen months for fuel, light, and power was $7,250; in comparison, his private secretary and aide-de-camp each earned $10,000. A footnote explained: "Considered necessary. Previously underestimated."[48] Although CLP focused on industrial consumers, during the post-1945 period electricity became more integrated into everyday urban life as the system to generate and distribute electricity was significantly extended. The 1954 CLP annual report noted "the considerable expansion of air conditioning" and predicted that "this form of cooling will replace fans as surely as the latter replaced punkas of old." By 1956 the company noted: "Air-conditioning is becoming increasingly popular, and is now a substantial factor in our load during the summer months."[49]

Electricity was in some ways more ordinary, less remarked upon, than it had been in its early years, yet it was combined with other technologies and used in ways that had a more thorough and far-reaching effect on the making of post-1945 Hong Kong than in the pre-1941 colony. Ice manufacturing demonstrated how electricity, combined with other technologies, transformed the city and the lives of many of its inhabitants. In the early years of the colony, the only significant source of ice was glacial and sea ice imported from Alaska. The first ice storehouse was established on Ice House Street, at the corner of Queens Road Central, in the late 1840s. Ice was imported from the United States by the Tudor Ice Co. Annual shipments failed only during the U.S. Civil War, when a ship carrying ice was captured and burned by Confederates. Ice-making began in the colony in 1874.[50]

Electricity made large-scale ice manufacturing routine and transformed a luxury into a commodity. Food manufacturer Dairy Farm's popular ice cream, sold by bicycle vendors throughout the colony, was a visible manifestation of a new consumer product made possible by the widespread adoption of electricity. After 1945, plentiful electricity prompted the building of new ice-making plants throughout the colony. With government help in building connecting infrastructure, the ice transformed the local fishing industry. Dairy Farm illustrated one of the new recombinations that electricity and the internal combustion engine made possible as well as the way that a private company could benefit from government subsidies: "The Company managed to get the government to build a three-lane closed, reversible conveyor, 170 feet long on reclaimed land that had become Victoria Park. This debouched at a new wharf built at the typhoon anchorage, where the boats now received the ice via a clattering mechanical device which took the block from the conveyor, dropping them into holds."[51]

Note that Dairy Farm was able to get the government to build infrastructure that would benefit the company and its customers. In addition to the 170-foot-long mechanical conveyor, perhaps the wharf and certainly the typhoon anchorage were paid for by the government. The government built the all-important supporting infrastructure that allowed for the combination of two new technologies, diesel engines and manufactured ice. This allowed for the benefits of the ice to be more economically shared than would have been the case if the ice had to be ferried by motor- or human-powered vehicles. By the mid-1980s, Dairy Farm owned three ice-making plants, two of which it claimed were the largest of their sort in the world, making the company "probably the largest single producer of ice in the world." Some 85 percent of that ice went to the fishing fleet.[52]

The availability of ice, coupled with the newfound availability of engines that enabled fishing boats to travel more reliably and over greater distances, had a far-reaching effect. Electricity and the internal combustion engine, two fundamental technologies, combined to radically reshape a traditional industry—fishing—that was an important part of Hong Kong's indigenous economy. Sailboats with little or no refrigeration were replaced by vessels that could range farther offshore and store their catch. This resulted in more income for fishermen and increased amounts of seafood, a source of protein, for both the local and export markets—but also in overfishing.

RESOURCES

Hong Kong's small size and unpredictable relationship with neighboring China prompted ongoing concerns about securing adequate sources of food, water, and fuel. Food was mostly imported from China, political tensions notwithstanding, although imports from Australia and elsewhere (mostly for consumption by expatriates) were notable. The need for significant quantities of uninterrupted and reliable water supplies increased as the population grew in the 1950s and 1960s. Water rationing was imposed for many years during the drier months.

Fuel cost, quality, and availability were recurring issues. For the electricity supply companies, until CLP secured supplies of nuclear- and hydro-generated electricity from mainland China in 1994, fossil fuels had to be burned for the company to profitably generate electricity. Anxiety over fossil fuel resources and the quest for economical, reliable supplies of coal and oil underpinned many of CLP's actions from its inception. Initially, the company apparently relied on supplies from Japan and Australia. At the annual shareholders' meeting in April 1907, founder Robert Shewan noted:

> The cost of coal was a little lower [in 1906] than during 1905, but it was still much above the normal figure, while on the other hand the quality is worse than ever, so much so much so that we have now about decided to abandon Japanese in favor of Australian coal. On the other hand, experiments with a diesel oil engine has [sic] so far resulted very satisfactorily with a substantial saving by the use of oil, compared with burning coal.[53]

Note Shewan's concern with managing the expense in what was the company's largest variable cost, fuel, and the need to ensure high quality. Also noteworthy are the early experiments with substituting diesel oil for coal as a fuel, a reminder both of the importance of managing costs and the fact that electricity-generating technology and techniques were still unstable. Left unsaid is the possibility that Japanese supplies were disrupted, and quality suffered during this period as a result of the 1904–1905 Russo-Japanese War. In the 1920s and 1930s, HKE obtained most of its coal from North China and mixed this with coal dust from the Hongay mines in Indochina.[54] A 1931 article on CLP noted that the plant used "a mixture of bituminous and semi-anthracite dust obtained from Japan and French Indo-China."[55]

Supplies of coal were insurance in a troubled world. At the December 1938 annual general meeting, less than two months after Japan's Imperial Army had captured Guangdong and placed its troops on the Hong Kong border, CLP Chairman J. P. Braga reassured shareholders that "the watchful policy of maintaining a sufficient supply of fuel has not been relaxed." He noted that the company had invested in "a considerable extension" in its coal storage ground. "The tying up of capital in coal stock is one of the safest forms of insurance for the special nature of your business."[56] Coal prices doubled between 1940 and 1941 as the war approached, and a 10 percent fuel surcharge was introduced in 1941.

The Japanese were unable to secure sufficient quantities of fuel to generate adequate supplies of electricity during their occupation of Hong Kong. There were increasingly severe electricity cuts during the Japanese occupation. At the end of the war, rice husks and firewood were used for fuel. Even after the British resumption of sovereignty, CLP continued for a time to rely on firewood. Kadoorie later wrote of the "hundreds of women continuously pushing logs of wood into the furnace" as the company burned up to 400 tons of wood a day, "which, thanks to the efforts of the Royal Air Force and fuel control, we were able to obtain."[57]

There are three points to note. First, coal was simply unavailable. Regular supply systems had collapsed during the war and had yet to be reestablished. Second, CLP's chairman chose to highlight the contrast between the technological modernity inherent in the 200,000-lb. boiler, which had the world's largest moving coal chain grate for fueling the facility, with the reality that normal operations had broken down to such an extent that hundreds of women were now pushing logs into the boiler. Third, the role of the Royal Air Force is yet another reminder of the symbiotic relationship between the military part of the colonial state and this private company, a relationship that attests to the centrality of electricity. In the immediate postwar period, the government took responsibility for distributing coal.[58]

The uncertainty about the cost and availability of fuel to generate electricity made the post-1945 decades a challenging time for CLP. The 1947 shareholders' report noted that new generating units could also burn oil and stated that coal supplies were adequate. The document lamented that the coal's quality was "very poor as compared with 1941 shipments, and the price still remains extremely high." The average cost in 1941 was $32 a ton, but $97 a ton in 1947. Coal appears to have been purchased from a variety of sources. In the mid-1950s, the Green Island Cement Co., located next to CLP's Hok Un generating station, bought coal from Japan, China, Indochina, Indonesia, India, and elsewhere.[59] This concern with coal was widespread in the colony. Writing in 1976, journalist Richard Hughes noted sardonically in the opening paragraphs of his book *Borrowed Place, Borrowed Time* that the colony "lacks coal, oil, and all natural resources save granite, sand, fish and *homo sapiens*."[60]

"DIE-HARD REACTIONARY" IN THE COLONIAL STATE

Experiments with oil-fired generation succeeded "beyond expectations," claimed the 1947 shareholders report, and CLP produced increasing amounts of electricity using oil. Oil prices, however, roughly doubled in the early 1950s, reaching $180 a ton in 1952, making predictable electricity pricing difficult; they fluctuated below the $180 level for the rest of the 1950s, reaching a peak in 1957 when the price averaged $167 a ton.[61] Oil prices fell throughout the 1960s to just $99 a ton when the Tsing Yi joint venture with Esso began operations.

The introduction of Esso as a partner secured the supply of oil and facilitated another tenfold increase in generating capacity from 1964 to 1982. But the rapid increases in the price of oil in the 1970s changed the economics of oil-fired electricity. The price CLP paid per ton of oil increased more than twelvefold from $99 in 1969 to $1,355 in September 1981.[62] Even the higher price could not guarantee availability. Hong Kong, and CLP, suffered through another round of resource anxiety after the 1979 oil shock, when prices increased sharply and supplies tightened following the Iranian revolution. Hong Kong's two electricity companies consumed more than half of the colony's oil and they depended on oil for most of their fuel needs.

Any disruption in oil supplies would have constituted a significant challenge for the city. In December 1980 and March 1981 two oil tankers, the *Straits Dahlia* and the *Seabreeze*, with a combined cargo of 197,000 tonnes, dropped anchor off Hong Kong and took on the role of a strategic oil reserve. They remained in place until the end of 1981, when land tanks on Lamma became available.[63]

The higher price of oil pushed CLP back to coal. Attempts to secure a reliable supply of an alternate resource in the form of coal intensified following the 1973–1974 OPEC oil embargo. In May 1977, CLP discussed the feasibility of purchasing coal from China with China Resources. In June, China Resources "indicated definite interest but made it clear that actual commitments would be deferred pending consideration by Peking." This had an imperial dimension, since Britain also needed coal. Securing coal supplies both for Hong Kong and for Britain was for Kadoorie a recurring topic of discussion with the most senior British officials.

In March and again in July 1977, Kadoorie met with Prime Minister James Callaghan to discuss the issue. In return for coal, Kadoorie hoped to secure contracts for British equipment suppliers to modernize the Chinese mining industry. At the July meeting, the prime minister "again showed interest in the purchase of Chinese coal for British equipment."[64] Kadoorie visited Beijing in May 1978, seemingly the first high-profile visit by a non-Chinese Hong Kong businessman, as the PRC was embarking on economic reform and opening under Deng Xiaoping.[65] Coal to fuel Castle Peak in its first decade of operation came from Australia, the PRC, South Africa, the United Sates, and Indonesia.[66]

Kadoorie remained fascinated with the possibilities of nuclear-generated electricity even after Hong Kong decided against building a nuclear reactor. In early 1981 CLP and the Guangdong Power Co. submitted a study on a nuclear power station.[67] In late 1982 the PRC government announced that it would construct a nuclear power plant at Daya Bay. CLP agreed to invest 25 percent of the equity in the plant and to buy 70 percent of the facility's output for the first twenty years of operation. Commercial generation at the two 985 megawatt units began in February and May 1994.[68] The Daya Bay project was China's first civilian nuclear facility and the country's largest foreign investment. Nuclear power caused Kadoorie to rethink the nature of state involvement in the economy. The capital investment required to build a dual-purpose nuclear-powered desalination and electricity-generating facility could justify a state-private joint venture, if "adopted by Government and paid for with taxpayer's money."[69]

The colonial government's struggles to secure reliable long-term water sources mirror those that CLP undertook for its fuel needs. Water was in short supply in Hong Kong in the 1950s and the first half of the 1960s. Anxieties about water supplies were much more serious than those about coal or oil. In 1963 a drought worsened an already brittle situation. Water use was limited to a few hours every four days. The government's review of 1963 noted "the remarkable spectacle of the great modern city of Hong Kong engaged in a desperate struggle reminiscent of earlier times and of more primitive societies—a struggle for water."[70] Water supply, like electricity supply, had political elements. Governor Robert Black signed an agreement with Guangdong authorities in 1960 for the province to supply the colony with water. At the June 1962 meeting to consider CLP's merger with HKE, shareholder D. Engel protested against nationalization and warned, "We should also soon have to have 'Red' electric current, as we now have to accept 'red' [sic] water from China."[71] An agreement for Shenzhen to supply Hong Kong with additional water, which relieved the colony's acute shortage, began in 1964, the same year that the SoCA was signed. These two agreements provided new sources of supply for two critical elements of mid-twentieth century urban life, piped water and adequate supplies of affordable electricity.

ELECTRICITY AND THE IDEA OF THE CITY

Cities are shaped both physically and symbolically. The interplay between the literal and the conceived makes up the idea of the city. Electricity was at the heart of the twentieth-century idea of urbanity and of urban experiences. Electric light and power amplified and extended urban possibilities by providing the current for lights, elevators, escalators, trams, air conditioners, televisions, ice-manufacturing plants, and much more. Twentieth-century Hong Kong was a self-consciously and self-identified modern, urban colony, one with its population concentrated in relatively small areas of Hong Kong Island and Kowloon. The colony's expanding electricity network, and the increasingly wide range of lighting and power applications, contributed to the sense that Hong Kong was modern. Electricity powered and lit the lives of increasing numbers of people in post-1945 Hong Kong, with a notable surge during the 1964–1973 period. There was a strong, ongoing interplay between electricity and a sense of living in a modern, urban colony.

The electric-powered lights of New York City's Times Square, Tokyo's Ginza district, and Tsim Sha Tsui's Nathan Road epitomized the visual element of the twentieth-century urban experience. Visual images helped underscore to contemporaries the role that CLP had played in bringing forth a new kind of colony in the 1950s and 1960s. During the mid- and late 1950s, CLP's annual shareholder reports featured pictures of electric street lighting, a neon-lit street in Kowloon, decorative lighting in Kowloon during the Mid-Autumn Festival, an oversaturated night picture of the Fifteenth Hong Kong Products Expo ("Hong Kong people using Hong Kong products"), and an illuminated picture of the Tsim Sha Tsui clock tower. The 1958 annual report showed Governor Robert Black switching on two new 30-megawatt generators at CLP's Hok Un generating station. CLP symbolized power—political, economic, and social. It also provided light—the light of enlightenment, but also the light of enjoyment and consumption and a vibrant nightlife.[72]

Electric light was often cited as a marker of modernity, one that distinguished twentieth-century life from that marked by the dim and dusky lights of the nineteenth century. Lawrence Kadoorie underscored the role of electricity in nurturing urban modernity. He looked back to a premodern Kowloon that was "lightless by night," its "narrow pathways.... lighted by kerosene lamps" and "narrow granite pillars surmounted by oil lanterns, which shed their dim light for the few passers-by." Kadoorie contrasted this with the steel-and-concrete electricity-generating station he was opening at Hok Un, one that was equipped with the latest model of Carrier air conditioning and a modern loudspeaker intercommunication system, as well as an air-raid shelter. The last notwithstanding, Kadoorie proclaimed that it was a building "designed with an optimistic view to the future," one that symbolized faith "in the greater progress and development of the Colony of Hong Kong."[73]

Electricity companies explicitly contrasted their product with what they derided as dangerous and unhygienic preexisting materials and practices. CLP emphasized problems with kerosene, notably fires, and equated older technology with dirty, primitive mores. A company-sponsored history quoted a resident of To Kwa Wan who remembered that the International Cinema there had had electric fans, not air conditioning, and that "the place was very dirty and seed shells were everywhere on the floor." The movie theater later installed electric-powered air conditioning. His father bought an air conditioner for the family's flat, partly to filter out the coal smoke from CLP's nearby Hok Un generating station: "Life became relatively comfortable after that."[74]

Power mattered as much as light. Ever taller skyscrapers were only possible because of electric-powered elevators. Electric-powered trams displaced horse-drawn trams and carriages and, in Hong Kong, rickshaws. Cool, smokeless, and convenient, like an obedient servant waiting for instructions that were given with the flick of a switch—this was electricity's promise. CLP and HKE reinforced the idea that their systems were exemplars of modernity.

HKE had the more explicit narrative. The company's 1974 annual report proclaimed: "Go Modern. Go Electric." Electricity is "safe, clean, convenient." For water heaters: "No ignition required. No fumes given out." For electric stoves: "No flames, no smell, no dirt." Electric hot water in the kitchen: "constant hot water, no need for ventilation."[75] Safety. Cleanliness. Convenience. Modernity. Electricity.

This was Hong Kong's aspiration in the 1960s and 1970s. More and more it was a reality, for the significant expansion of generating and distribution capacity at both CLP and HKE enabled hundreds of thousands of new consumers to receive electricity for the first time. In June 1982, Kadoorie commemorated the addition of CLP's one millionth residential consumer by presenting the Wong family of the newly completed Shun Tin Estate with a plaque and a videotape recorder.[76]

The electricity companies used visual images of modern technology to underscore the contributions they were making to Hong Kong's modernization. These images imprinted and reinforced the idea that the electricity supply companies were importers and transmitters of a modern, urban technological culture. Photographs in the companies' annual reports of technology including VHF-radio-equipped vehicles and modern computer rooms were one way in which a specific capital- and technology-driven vision of modernity was highlighted.

One of the most enduring uses of Hong Kong's urban lighting was in celebrating global events, especially those of the British Empire. HKE's first-ever light show was held in Central's Statue Square to mark the signing of the Versailles Treaty in 1919.[77] Kadoorie told shareholders in 1935 that the company had commemorated King George's Silver Jubilee with one week of illuminations in parts of the New Territories that were "very effective and much appreciated by the elders and the populace."[78] To commemorate the coronation of Queen Elizabeth II in 1953, CLP's Argyle Street headquarters was bedecked with heraldic arms. Underneath the arms, the script read: "God Save the Queen." The Hok Un station was floodlit for the coronation celebrations in June. The lighting installation was then made permanent.[79]

A similar exaltation of this version of technocentric modernity was provided by contrasting images of traditional Hong Kong with the electricity transmission network. The steel pylons that supported the cables carrying electricity across Hong Kong were popular visual images, especially when contrasted with pictures that ostensibly showed older, more traditional ways of life. HKE's new Ap Lei Chau generating facility and the high-voltage overhead utility wires that transmitted its electricity served as the backdrop to an annual report photo dominated by fishing trawlers, symbols of Hong

Kong's maritime tradition. In fact, by the time this photo was taken in 1973, most of those fishing trawlers were equipped with diesel engines.[80]

The limits of Hong Kong's electric-powered urbanity can be seen by contrasting the colony with cities such as Berlin and Chicago. Hong Kong had no artists like László Moholy-Nagy, who celebrated the possibilities of electric light with his renowned 1930 *Lichtrequisit einer elektrischen Bühne* (light prop for an electric stage).[81] The Hungarian-born Moholy-Nagy was for a time part of the Bauhaus movement in Berlin. He was among the most successful explorers of the possibilities of electricity among this group of artists and architects who developed new architecture and new ways of living by using electricity in combination with modern materials like steel, glass, and cement.

Although Hong Kong had no Bauhaus movement, CLP's Argyle Street headquarters was one of the colony's small number of Bauhaus-influenced buildings. Hong Kong also had no Frank Lloyd Wright, an architect whose innovative houses subtly incorporated electric light.[82] Although Wright's vision was quite different from that of Moholy-Nagy and the Bauhaus movement, with Wright favoring more organic forms and the use of wood, he, too, exploited the technological possibilities provided by the availability of large, durable pieces of glass and steel combined with inexpensive, reliable, and ubiquitous electricity when designing the spaces in which we live and work.

With the exception of neon signs, which explored the consciously futuristic possibilities of electricity, Hong Kong's creativity in electricity was mostly functional and solutions-oriented, not celebratory or visionary, let alone self-consciously artistic. Hong Kong's filmmaking industry boomed during the 1950s and 1960s. Movie production, which required significant amounts of electricity, is another sort of exception, one that used the technology without celebrating it.

Hong Kong's businesses did take advantage of the new forms of urban consumption that electricity enabled. Kowloon's Tsim Sha Tsui area, where the Peninsula Hotel was located, became an internationally renowned entertainment center in the 1950s and 1960s. The area's Miramar Hotel boasted of its large Broadway-style nightclub and was described as "the most modern and westernized entertainment place in the region," with correspondingly large electricity consumption. In the 1960s the large, brightly lit, and air-conditioned Ocean Terminal ushered in a new kind of shopping experience for Hong Kong and the region.[83]

"DIE-HARD REACTIONARY" IN THE COLONIAL STATE

CLP also used photographs to show how the modernizing force of its electricity was changing Hong Kong. Workers building the Hok Un extension held a traditional Chinese New Year's ceremony against a backdrop of three large cable spools, each almost twice the height of the men in the picture. Three large hanging scrolls depicting Chinese deities hung on the center spool, which functioned as a temple wall.[84] The 1960 annual report, as mentioned earlier, showed a woman wearing a traditional hat handpoling a sampan. Her freight consisted solely of a cash receipting machine—an emblem of modernity—that was en route to Tai O, Lantau, for the bimonthly opening of CLP's bill payments office.[85]

The photos could be used to demonstrate an alternate vision of urban modernity. Photos in the 1959 annual report of stolen, damaged, and destroyed bits of CLP's electricity network complemented Kadoorie's testimony to the ESCC, a visual warning about what could happen if electricity companies were forced to bend to populist pressures such as the nationalization campaign. For Kadoorie, nationalization appeared to be synonymous with decay and ruin; the analogy for him was the destruction of private companies in Shanghai after the 1949 Communist takeover. These photos include a dramatic closeup of a transformer destroyed by lightning; cables and a joint that had been exposed by road subsidence during heavy rains in June 1959; and a cable in Boundary Street "looted of its armour and lead sheathing by thieves." The fourth and final picture is of a tidy open-air substation at Au Tau.[86] This was the choice that Hong Kong people needed to make, the company seemed to be suggesting: Did they want a city where movie theater floors were littered with seed husks; where lightning and storms destroyed equipment; where thieves looted equipment that should have brought light and power to ordinary citizens? Or did they want a city of tidy, air-conditioned cinemas and well-lit streets powered and illuminated by a company that had VHF radio-phone-equipped repair trucks, gleaming generating and transmission equipment, and modern computer systems? The answer was self-evident.

THE WORKING PARTY ON THE ELECTRIFICATION OF SQUATTER AREAS

Many of the elements that identified electricity with urban modernity came together in 1975–1976 with a pilot project to electrify squatter settlements in villages below Lion Rock.[87] The squatter electrification project marks a notable moment in the emergence of the infrastructure state—one that defines building physical infrastructure as a primary function—and the development of mutually reinforcing structures of technical and administrative power. This project reflected and in turn reinforced the emergence of a more confident and assertive colonial and corporate state. The pilot project was a kind of bricolage, one that awkwardly stitched together the precision of electricity supply with the informal, irregular nature of squatter settlements. It represented an example of imposed modernity, one in which a state that took increasing responsibility to provide infrastructure was forced to improvise and compromise. To implement the program, the government had to accept that squatters would remain in large numbers for many years to come and that it should take responsibility for improving their lives.[88]

By 1975, electricity was officially regarded as a "necessity" of modern life. The pilot project was designed as a first step in plans to supply electricity from CLP and HKE to Hong Kong's more than 270,000 squatters. The government was "anxious" to improve squatters lives by making safer, and less expensive, electricity available in virtually every home, even if the dwellings were not legally compliant. By 1975, most squatters had access to electricity. This was typically provided by private operators who stole the electricity from CLP or HKE or produced it using their own generators. In either case, the electricity was distributed in a way that did not meet safety requirements and was therefore illegal. This situation, the report on the pilot project noted, threatened the integrity and safety of the electricity system and also "involved a serious loss of revenues" for the supply companies.[89]

Secretary for Housing I. M. Lightbody established a Working Party on Electrification of Squatter Areas in November 1975 with the aim of finding a pilot area in which to install a legal electricity supply and to overcome the practical problems involved.[90] The group knew from the outset that it faced a range of legal, financial, safety, material, and technical obstacles. The most important issues centered on the electricity supply companies' legal liability. Laws were accordingly amended in 1976 to allow supply to less well-constructed dwellings. This gave CLP confidence that it would be paid for electricity supply and that it could control that supply. A key part of the scheme centered on the installation of consumer meters in special centralized locations built by the company. Consumers were responsible for the wiring that connected the meter to their dwellings. Safety was ensured by the installation of a circuit breaker at each meter in the main pillar as well as an earth leakage circuit breaker in the dwelling itself. Most of the cost was borne by CLP. Consumers were given the ability to select their own wiring contractor, with contractor choice widened as a result of protests by villagers who wanted to choose from an array of different prices and standards of workmanship.[91]

Most squatters lived in CLP's service area, so it was appropriate for CLP to run the pilot project. The working party that was established in November 1975 included two CLP staff among its seven members, Commercial Engineer Colin Steele and Transmission and Distribution Engineer R. A. Pedder. HKE did not have any members of the working group, but the company's G. A. Hume served on the working party's legal subcommittee.[92]

The pilot area was located, fittingly enough, under Lion Rock.[93] The 35-hectare area contained more than 1,150 structures housing 2,000 families. It included about 9,000 people. The goal was to provide electricity to every squatter hut in the area, "no matter how makeshift," although certain safety minimums were enforced. To receive electricity, a structure was required to be waterproof, covered, and enclosed on four sides, and "steady and secure in adverse weather conditions." These requirements were, it was said, simply designed to protect inhabitants against the possibility of electric shocks or fires "and were unrelated to any assessment of whether the structure was considered fit or safe for human habitation." City District officers, working through existing local mutual aid organizations and those that the government helped establish for the purposes of the project, advised squatters of the regulations. "Villagers subsequently made the necessary repairs in compliance with the minimum standards."[94]

The process was one in which villagers who wanted electricity were brought into an administrative net. In addition to the requirement that dwellings be certified to a minimum standard, inhabitants had to be able to receive electricity bills. Anyone wanting electricity thus needed to arrange for postal delivery.[95] Conflicts occurred between the rule-bound administrative approaches adopted on one side by CLP and government officials and, on the other, by villagers' desire for flexibility (and lower costs) in hiring contractors to install the wires that led from the central pillars into the consumers' homes. The working party's report contended that most contractors "had little knowledge of electrical installation work, and opportunities for irregularities would arise if the local organization were to engage its own contractors to inspect consumers' installations." The companies and government compromised and allowed a wide range of contractors to take part to avoid further delays to the project, although the Independent Commission Against Corruption (ICAC) was brought in to minimize irregularities in the bidding process.

Inspections were instituted to validate contractors' work after the working party decided that it was "essential on safety grounds" for an "independent competent body" to inspect consumer installations before they were connected to the electrical mains supply. It was first proposed that the government should inspect the wiring, but this idea was dismissed because "Government felt it would be wrong in principle for it to act as an intermediary in what was essentially a purely contractual relationship between a utility company and its consumers." This was an example of the Hong Kong government shirking what would in most places be considered normal state responsibilities to ensure safety under the guise of pursuing its limited-government principles.

Ultimately CLP conducted the inspections without charge, "thus overcoming what might otherwise have been a substantial stumbling block." This compromise brought CLP officials into private dwellings, even as the company retreated from the stricter oversight of contractors that had initially been proposed.[96]

Government action was evident throughout. The Royal Air Force helped transport materials to hillside sites that lacked roads. The Hong Kong Police, the Squatter Control Division of the Housing Department, and CLP staff dealt with security matters, site clearance, discussions with private generator operators, and coordination of the clearance of some unregistered squatter structures.[97] The project included an evaluation of firefighting capabilities. These were found to be substandard, and plans were underway, at the time of the publication of the working party's report, for improvement of firefighting facilities in the area.[98]

Taken at face value, this improvement appears paradoxical, since fire danger was reduced with the introduction of regulated electricity supply and CLP oversight. Indeed, reducing fire danger was regularly cited as a reason for regularizing electricity supplies. Both private generators and the kerosene lamps and stoves that up to 20 percent of villagers used in one village in the area posed a greater danger of fire. Yet the extension of firefighting capabilities as an ancillary part of electrification shows that the broader aim of the project was nothing less than to reconfigure the colonial technical infrastructure.

The first switch-on ceremony in a squatter village was held on June 11, 1976. Yip Wai Ying, a fourteen-year-old school girl, turned on the electricity in Upper Lion Rock Village. Village Mutual Aid Committee Chairman Hou Tin-sang expressed the gratitude of villagers who would no longer have "to live in darkness or under the threat of fires." Lawrence Kadoorie also spoke at the ceremony. Unlike his remarks at Hok Un, Kadoorie did not speak in grand historical terms. Instead, he simply "expressed confidence that the pilot scheme could, with goodwill and cooperation, be followed in due course in other areas."[99]

Kadoorie's understated remarks notwithstanding, the successful installation of electricity throughout the colony was indeed historic, part of a pattern in which the city and its people were remade in a new image. Squatters would increasingly have access to the electricity that was now regarded as the right of every person in an ever more prosperous colony thanks to actions by its more proactive and self-confident administrative and business leaders.

Chapter Seven

"INTELLIGENT ANTICIPATION" FOR "1997 AND ALL THAT"

1974–1982

Margaret Thatcher's widely televised fall down the steps of Beijing's Great Hall of the People in Beijing on Friday, September 24, 1982, epitomized the troubles of a difficult day for her.

That morning, paramount Chinese leader Deng Xiaoping had bluntly told the prime minister that the resumption of full Chinese sovereignty and administration over the British Crown Colony of Hong Kong in 1997 was nonnegotiable, effectively ending the British attempt to convince the Chinese to agree to continue British rule in a more limited form. It was not at all how Thatcher, buoyed by Britain's recent military victory in the Falklands, had envisioned the negotiations, but her visit to Beijing and Hong Kong set the stage for the final British retreat from the colony on June 30, 1997.[1]

Thatcher's fall and her meetings in Beijing also symbolized a dramatic reversal of historical fortunes. Britain's first envoy to China, George Macartney, had by his account famously refused to kowtow to the Qianlong Emperor during his embassy to Peking in 1793. The intervening two centuries had seen Britain defeat China in the two Opium Wars and impose what the Chinese regarded as humiliating extraterritorial treaties, the terms of which included Qing cession in perpetuity of Hong Kong Island in 1841 and the Kowloon peninsula south of Boundary Street in 1860 and a ninety-nine-year lease on the New Territories in 1898. With the expiration of the New Territories lease only fifteen years away, Deng and China's other leaders were determined to erase the colonial legacy of unequal treaties embodied in British rule over Hong Kong.

Four days after Thatcher met Deng, the British prime minister inaugurated CLP's Castle Peak electricity generation plant, located in the western part of Hong Kong's New Territories. The construction of the station, and an even larger second part of the facility that was in progress when Thatcher visited at a time of growing anxiety over the future of Hong Kong after 1997, testifies both to the importance of that most material form of industrial power—centrally generated electrical power—in the making of modern Hong Kong and to Kadoorie's long-term investment philosophy. The decision to support Castle Peak's construction, which comprised the largest export order ever received by the British manufacturing industry, had been made at the highest levels of the British government with the direct involvement of Thatcher and her predecessor as prime minister, James Callaghan.[2]

Pragmatic decisions concerning electricity encouraged political compromise on Thatcher's part during her discussions in Beijing with Premier Zhao Ziyang, whom she urged to select British contractors for the proposed Daya Bay nuclear plant that was being championed by Lawrence Kadoorie; China's need for electricity in turn made it amenable to working with Britain on energy projects. Electricity transmission lines crossed the border, one of the first physical linkages between the colony and the PRC at a time when British maps showed blank space north of the Hong Kong border. Electricity, as a foundational technology required for modern economic activity, was a potent political and diplomatic force in the two decades preceding Britain's retreat from Hong Kong.

Electricity affords us a new perspective on the worlds of elite diplomatic politics and international finance. The Castle Peak project was a political as well as a business choice, a bet on the future of Hong Kong. Its political impact proved far-reaching. The role of British industry in building the facility and British banks and financial institutions in providing the funds for the electricity-generating station in turn ensured that the colony of Hong Kong itself benefited from continuing political attention in the run-up to the 1997 political transition, as evidenced both by Thatcher's inauguration of Castle Peak in September 1982 and the opening of the facility's second phase by Prince Philip in 1986.

Castle Peak also helped Kadoorie put the "China" back into China Light & Power. In 1979, the company had started selling electricity to China for the first time since it had left Canton in 1909. Castle Peak's new generating capacity ensured that CLP would have ample reserves to meet the electricity requirements of neighboring Guangdong. These endeavors facilitated Kadoorie's successful championing of China's first commercial nuclear power plant, at Daya Bay, a project that was approved shortly after Thatcher returned to Beijing in 1984 to sign the agreement with China that would mark the end of British rule of the colony in 1997. In 1985, shortly after the Daya Bay project was finalized, Deng Xiaoping met with Kadoorie, which signaled the project's—and Kadoorie's—importance.

How Kadoorie organized the CLP electricity system in anticipation of the expected end of Hong Kong's status as a colony in 1997 reflected his assessment of broader political and financial concerns. His analysis gives us insight into Hong Kong's future as it was viewed in the colony, as well as in China and Britain, in the mid- and late 1970s and early 1980s, some fifteen to twenty-five years before the expiration of the New Territories lease. Kadoorie, one of the wealthiest and most important business leaders in Asia, used the construction of the Castle Peak power station both as a way of ensuring continued close British involvement with Hong Kong and to guarantee that Hong Kong would become increasingly useful to China. He was fond of saying that Hong Kong was founded on a "three-legged stool." Paraphrasing Kadoorie, the *Financial Times* wrote: "The first leg is that Hong Kong must show that it is of some use to Britain; the second is that it must show that it is of use to China; and the third, that it must show it can give better jobs to the more sophisticated and educated younger generation in Hong Kong."[3]

It is also important to note that since the 1950s Kadoorie had been certain that in 1997 China would resume sovereignty over the New Territories at least, and probably the entire colony. Many people did not believe that 1997 was a terminal date; Kadoorie gave no indication that he was one of them.

To what extent did Kadoorie's plans to engage with a more economically active China align with those of Thatcher and British industry at a time when Thatcher was redrawing the boundaries of the state and limiting direct government intervention in industry through a program of denationalization and deregulation? How did the construction of the Castle Peak facility address contemporary anxieties about resources? Finally, how did Castle Peak and its network fit into a larger constellation that included Hong Kong, Britain, China, and the United States? The answers to these questions enable us to better understand the growth of Hong Kong's economy and changes in its politics and society in the two decades before the end of British sovereignty.

Most histories of Hong Kong have emphasized politics, and political elites, with little attention paid to the changing commercial milieu within which politics was played out. The inauguration of Castle Peak was not only a business venture and a practical solution to the problem of meeting the colony's energy needs. Beyond this, it was also an episode in Sino-British relations, one which highlights both the anxieties and the possibilities that Hong Kong's looming handover to China presented.

"1997 AND ALL THAT": A SHOW WINDOW FOR THE WEST

Lawrence Kadoorie in the mid-1970s set forth an unusually candid series of reflections in three successive versions of an essay entitled "1997 and All That." It was prepared for directors at PEPCO, the joint venture between Esso and CLP that built and operated the Tsing Yi oil-fired electricity-generating station, a facility that by the mid-1970s provided the bulk of CLP's electricity. Esso's dominant position in the joint venture coupled with Hong Kong government restrictions on CLP's capital spending and its ability to pay shareholders dividends meant that CLP could not make major investment decisions alone. Kadoorie needed Esso's financial support and thus was forced to argue his case in more detail than he would have otherwise. These wide-ranging and previously unpublished pieces provide unusual insight into Kadoorie's hopes and fears at a time of increasingly large-scale energy use and Hong Kong's impending political transition. They also reflect the impact of sharply increased energy prices beginning in 1973 and heightened anxiety about energy supply.

Kadoorie's purpose in writing "1997 and All That" was to analyze the need for new electrical generating and transmission capacity in light of the uncertainty about what rule by the PRC would mean. The essays are an essential part of a process that culminated in the decision in late 1976 to construct Castle Peak. The facility was one of Kadoorie's most important achievements. Castle Peak's two phases tripled the size of the electricity company, and the increased profits it generated established the Kadoories as one of Asia's wealthiest families.

The three versions of the essay, written over a period of more than two years, show the development of Kadoorie's thinking as he assessed the future of Hong Kong after 1997. The differing versions also enable us to understand his response to international events as they unfolded. When he completed the second and third versions of his essay, in 1975 and 1976, there was heightened concern that bordered on panic in international energy markets. Changes afoot in China and Britain, though as yet invisible to Kadoorie, would dramatically reconfigure the British and Chinese states. CLP would find itself the beneficiary of these changes, ones that in Britain gave a more central role to privately owned corporations, particularly to investor-owned electric utilities. Under the Thatcher administration, electric utilities would be sold to private investors and would regain the role as private corporations that they had lost with nationalization in the late 1940s.

Lawrence Kadoorie completed the first of three versions of "1997 and All That" in October 1973. The title, its air of jaunty insouciance notwithstanding, hints at the importance of the work, for it marks one of the few instances that Kadoorie used anything other than a factual description as a title for a report or speech.

He had to make the case for significant new investments at a time of gathering political uncertainty in Hong Kong, to executives employed by one of the world's largest oil companies, one whose operations spanned the globe. He was thus forced to articulate his views on Hong Kong to an audience that had no particular allegiance to or knowledge of Hong Kong, let alone a deep or sympathetic understanding of China. He used the full force of his intellect to argue in favor of a sizeable investment, one that would triple CLP's generating capacity over the next decade while enabling the company to recoup its capital well before 1997.

The importance of these essays is visually evident from the care lavished on their physical presentation. Although "1997 and All That" was produced in limited numbers for an internal audience of the most senior executives, its format sets it apart from other corporate documents in the Kadoorie archives at the Hong Kong Heritage Project. The copy of the October 1973 version is distinguished from the hundreds of brown folders that hold most of the CLP archival material at the HKHP by its blue cardboard cover and its embossed "Confidential" label. "No. 2 of 10" appears on the title page along with Kadoorie's signature in blue ink. Addressed to "To the Directors CLP/PEPCO," the first page of text warns "NOT FOR PUBLICATION." (The first version of the essay was addressed to ten directors, the second to six.) After completing a first draft in October 1973, Kadoorie subsequently revised the study twice, in 1975 and 1976, making substantial and significant additions.

His increasingly lengthy essays all begin with an extensive historical discussion of the three treaties that form the basis for Britain's rule over Hong Kong. His sober summary of the colonial history of Hong Kong was made with the aim of assessing prospects for CLP after the expiry of the ninety-nine-year lease on Hong Kong's New Territories in 1997: "This study has been prepared with a view to ascertaining my colleagues [sic] thoughts on a subject of growing importance and in the hope of establishing a constructive program for the longer term protection of our respective companies and the shareholders whom we represent."[4]

Titling the work "1997 and All That" is almost certainly a wry reference to the popular British history *1066 and All That*. The best-selling book, serialized in *Punch* and first published in book form in 1930 as the Depression deepened the sense of imperial decline, was an irreverent look at two millennia of the history of the British Isles. The work's satirical but relentlessly optimistic gloss on British history tells the story of successive waves of invaders who conquered the island; despite, or perhaps because of, various humiliations imposed by the foreign conquerors, the country thrived. The authors of *1066 and All That* summarize the impact of the Norman conquest that occurred in that year: "The Norman Conquest was a Good Thing, as from this time onwards England stopped being conquered and thus was able to become top nation."[5]

Was Lawrence Kadoorie suggesting that the end of British rule would prove a good thing for Hong Kong? There is nothing in the historical record to suggest that Kadoorie welcomed the resumption of PRC sovereignty over Hong Kong. He wanted to position the company so that the impending political transition could be used to CLP's advantage by facilitating cross-border electricity sales as well as investment. "1997 and All That" suggests that Kadoorie was confident Hong Kong could find enough space to thrive as a kind of safe zone, or, in his words, "a neutral point of contact in a shrinking world where the civilisation of both East and West can associate to mutual advantage."[6] Kadoorie seems never to have expressed the sorts of doubts about the impact of Chinese rule in Hong Kong common among British and Chinese businessmen.

On the final page of the first "1997 and All That" study in 1973 Kadoorie had a single word, in large black capital letters, on an otherwise blank page: "THINK." Think he did: Kadoorie's aim in these essays was to consider how CLP could best take advantage of the shifting political and business climate.

A close reading of the three versions of this essay yields important understanding of how Lawrence Kadoorie was positioning himself with relation to Hong Kong, Britain, and China when all three were in the midst of or approaching a time of major transition. Within two years of the completion of the final version of Kadoorie's essay in 1976, Deng Xiaoping and Margaret Thatcher would assume leadership positions in their respective countries. Both would redraw long-held conceptions about the nature of their states, especially in the business and economic realms. In both countries, the late 1970s saw the start of historic shifts wherein the state retreated from significant parts of the economic and business spheres.

As a result of the actions Lawrence Kadoorie took after the completion of "1997 and All That," he and his Castle Peak power project would be significant beneficiaries of the historical reshaping underway in these two countries and in Hong Kong. Until the ESCC hearings in 1959 Kadoorie had been a vocal opponent of government involvement in his business. The SoCA showed that there were benefits to CLP in working with government. These essays show the continued evolution in his thinking. Kadoorie would draw on the increased resources of the modern state, especially in Britain, to enable his company to successfully embark on a period of unprecedented growth.

He thought it almost certain that the PRC would take over the New Territories, and probably all of Hong Kong, in 1997, if not earlier. As he worked through successive versions of "1997 and All That," Kadoorie was increasingly upbeat about what this change would mean for CLP. Chinese pragmatism would prevail, he believed. The end of British rule would open up the possibility of increased economic integration with China. China's economic development would require copious electricity. CLP and its cross-harbor counterpart, HKE, could provide the needed electric current, tangibly demonstrating Hong Kong's usefulness to the PRC and providing new growth opportunities for CLP.

In the 1973 version of "1997 and All That," Kadoorie cautioned CLP and PEPCO directors: "A change in the status of Hong Kong could take place at any time. It could be before 1997 and it could, conceivably but improbably, be after 1997."[7] In the 1975 version, Kadoorie warned: "The time is rapidly approaching when Directors of companies and others holding responsible positions in the Colony will have to inform their shareholders and the public of the relevance of the expiry of the New Territories Lease."[8]

The question of what would happen to Hong Kong when the New Territories' lease expired had long concerned the colony's leaders. As noted earlier, on a 1954 lecture tour of the United States, Hong Kong Governor Alexander Grantham had remarked, "1997 will be the fateful year, for in that year the lease of the New Territories runs out, and I could not conceive of any Chinese government of whatever complexion renewing the lease. Nor could I imagine the rump of the Colony—the island of Hong Kong and the tip of the Kowloon Peninsula—continuing to exist as a viable entity."[9]

What would 1997 mean? From the vantage point of 1970s Hong Kong, that answer must have seemed particularly perplexing. In the nearly eight decades since 1898, when the lease on Hong Kong's New Territories had taken effect, China had seen the end of the two-centuries-old Qing dynasty, the short-lived Republican period, Japanese invasion, civil war, Communist triumph, and, beginning in 1949, an unpredictable and largely closed PRC. A United Nations embargo on trade with China imposed in 1951, following the People's Republic entry into the Korean War, simultaneously disrupted Hong Kong's trade with China and opened up new opportunities for manufacturers in the colony.[10]

Well into the 1970s, China remained largely sealed off from the world. There was deadly infighting at the top levels of the Communist Party. In 1971, Mao's putative heir apparent Lin Biao died in a plane crash while fleeing the country. In 1973 Mao's wife Jiang Qing and other members of the so-called Gang of Four reignited the political struggles of the Cultural Revolution with a campaign against Premier Zhou Enlai. Mao's death in

September 1976, which quickly led to the downfall of the Gang of Four and made possible Deng Xiaoping's rise to power, occurred nine months after Kadoorie finished the third version of "1997 and All That."

Absorbing Hong Kong into the PRC was not a Chinese priority. Mao had been more concerned with recovering Taiwan than seizing Hong Kong. Long-time Premier Zhou Enlai summed up the policy in 1954 by saying that "the timing is still not yet ripe" for the People's Republic to take back Hong Kong.[11] "Hong Kong should serve as a useful port for our economy," Zhou said in 1957.

In 1960 Zhou said that the country had adopted a principle of "taking long-term views and taking full advantage" of Hong Kong. As explained by an official Hong Kong government publication issued on the fifteenth anniversary of PRC rule, "taking long-term views" meant that China would not immediately act to "resume" sovereignty. The idea of "taking full advantage" meant that China could benefit from Hong Kong's position as a capitalist city.[12] From the PRC's founding in 1949, there had never been an attempt by Beijing authorities to enforce a Berlin-style blockade, despite the fact that Hong Kong was militarily indefensible and could have quickly been forced to capitulate. Yet, as noted, China continued to provide Hong Kong with critical supplies of both water and food.

During the 1970s, China became more diplomatically engaged and assertive. The PRC won its United Nations seat back from Taiwan in 1971. Richard Nixon made his historic visit to Beijing in February 1972. The following month, Britain and China normalized diplomatic relations. The PRC mounted a campaign to convince other countries that the issues of Hong Kong, Taiwan, and Macao were internal affairs of China and could not be compared to a typical decolonization campaign that would lead to independence. An official Hong Kong government document notes that during "meetings with foreign guests, Zhou Enlai clearly stated that the future of Hong Kong must be determined and that China and Britain must hold talks on the lease due to expire."[13]

The official Chinese record of the period remains confused to this day. According to the official Hong Kong history of the development of "One Country, Two Systems," in May 1974 former British Prime Minister Edward Heath met Mao Zedong, and they agreed that "there should be a smooth handover of Hong Kong in 1997," although the account provides no further detail.[14] What purports to be a Chinese transcript of their talk shows that Mao was not eager to push for a change in the status quo. Instead, according to this account, Mao says of Hong Kong: "We won't discuss it at present. We shall consult together at the proper time about what we are going to do. This will be the business of the younger generation."[15]

The idea of "One Country, Two Systems," whereby Hong Kong would be granted a high degree of autonomy in running its affairs for fifty years after 1997, was not developed until 1979, according to the official Hong Kong account, and was not publicly used by Deng Xiaoping until the year he met Thatcher. The formula was initially intended to apply to Taiwan. During Deng's visit to the United States, according to the official Hong Kong government account, the Chinese leader "publicly declared that 'we will respect the realities and current systems there so long as Taiwan returns to the Motherland.' This was the first time a national leader had articulated the idea of 'One Country, Two Systems' which could be followed within the sphere of one nation." Deng, according to the official Hong Kong account, did not publicly use the phrase "One Country, Two Systems" until January 1982, little more than eight months before he met Thatcher.[16]

Kadoorie, in short, had very little basis for optimism in the mid-1970s. He knew that Zhou Enlai, and even Mao, had been pragmatic in their dealings with the colony. He might have known that in March 1973 Mao had appointed Deng Xiaoping as vice premier, initiating Deng's political rehabilitation. But the implications of Deng's rise to power would have been impossible to foresee. When Kadoorie wrote "1997 and All That," he could perhaps sense that the Chinese would become more assertive on the issue of Hong Kong, as China had in recent years lined up allies to support Beijing's contention that Taiwan, Hong Kong, and Macao were domestic Chinese concerns, not international ones. Yet he would have had little reason to believe that anything like the sort of autonomy promised in "One Country, Two Systems" would develop, since the Cultural Revolution was continuing as Kadoorie wrote his series of essays.

Whether Kadoorie could feel the political and diplomatic winds shifting or his habitual optimism and belief in Chinese pragmatism pushed him in that direction, he shrewdly used the historical record to make his case to Esso directors, one that suggested how a pragmatically minded Chinese leadership might be likely to rule Hong Kong. In the opening section of "1997 and All That," he gives a detailed summary of the various treaties that underlay British rule of Hong Kong, as well as specific Chinese responses to points of friction. Kadoorie paid particular attention to China's stance toward the Kowloon Walled City, a Chinese exclave that was not subject to colonial sovereignty. He concluded that although China did not recognize the treaties as valid and reserved the right to act as it wished, the country would most likely act reasonably, as it had done with the Kowloon Walled City.

In his "1997 and All That" essays, Kadoorie argued that a practical solution could be found to ensure Hong Kong's continued prosperity. Kadoorie thought that Hong Kong would be reabsorbed into China but that it would somehow continue to exist in a sort of undefined middle space, not a colony and certainly not independent, part of China yet in some vaguely framed way separate. Almost a decade before Deng Xiaoping publicly proposed that Hong Kong could become a Special Administrative Region, Kadoorie suggested that Hong Kong could be what he termed a Special Area:

> Looking forward some insist that Hong Kong as a British Colony makes confrontation with China on this issue inevitable. Personally, having been born in Hong Kong and as an old timer in the Far East, I am more optimistic.
>
> I see no "loss of face" in accepting as fact that with a population where only one person in one thousand is not Chinese Hong Kong is and must be a "Special Area."
>
> If and when it becomes expedient to recognise this view officially, anticipate no need for confrontation. Adjustments in foreign relationship can take place by negotiation and without undue friction.
>
> To my mind the Hong Kong Government's role will become more and more one of management and the British connection more and more that of a bridgehead or neutral and flexible meeting place between the ideologies of East and West.[17]

The Hong Kong government's role, in Kadoorie's view, would be to act as a technocratic manager. There would no longer be outright British administration, but a "British connection" would remain. In mid-1975, almost two years later, Kadoorie was even more buoyant: "I can see the growing importance of Hong Kong—both internationally and intercontinentally, as the free zone of China and as a neutral point of contact."[18] He repeated a concept he had articulated in 1973:

> In such an atmosphere there is no reason why Hong Kong should not remain as a show window for the West, a permitted society where, subject to China's veto, both foreign and Chinese residents can contribute practical international co-operation in Western culture and in the use of Western Technology, a neutral point of contact in a shrinking world where the civilisation of both East and West can associate to mutual advantage.[19]

Mao was still alive when Kadoorie wrote this version of the essay, so this assessment came from Kadoorie's analysis rather than any privileged information about China. The metaphor of Hong Kong as a "show window" conveys the idea of light and theatricality as well as of commerce. Kadoorie blended his understanding of the role the colony played in bringing western technology and ideas to China with his speculations about what Hong Kong might look like, if conceived of in a thoughtful way by the Chinese, after 1997. This allowed him to make the case for Esso to invest in Hong Kong on an unprecedented scale. As events unfolded, he was proven correct: Deng's program of pragmatic economic reform closely resembled Kadoorie's scenario. The outcome was even more favorable than he imagined, as CLP's assets were not nationalized or subject to a forced sale and China's economic opening provided a new source of demand.

Kadoorie did more than project a sense of unfounded optimism. He had thought his way into the future in a way that put him a decade or so ahead of most of his competitors in the colony. When Thatcher visited in 1982, the dominant mood among both the British and Chinese business community, or at least that part of it that met with the prime minister, was fear and caution.

Many people hoped that Britain could continue administering the territory. Hopewell's Gordon Wu warned Thatcher: "Chinese leaders might say that they wanted to maintain the status quo in Hong Kong but how good was their word? Their track record was terrible." Li & Fung's Victor Fung argued for "the continuation of the status quo indefinitely, with 25 or 50 years notice of termination." Fung wondered whether "China and Britain meant the same thing when they talked about maintaining the stability and prosperity of Hong Kong."[20]

"INTELLIGENT ANTICIPATION" FOR "1997 AND ALL THAT"

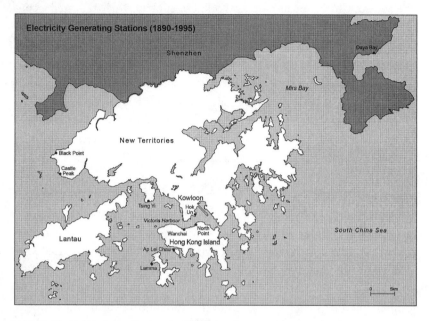

MAP 7.1. The Hok Un site was for decades CLP's only substantial electricity-generating station. Growth of some 20 percent annually in the 1950s meant that a new site was needed. Esso's 1964 investment financed a new plant on Tsing Yi. When Tsing Yi proved too small, Castle Peak and Black Point followed. The Daya Bay plant, which started operations the week after Lawrence Kadoorie's death in 1993, was China's first commercial nuclear reactor and the country's largest foreign-funded investment. Coupled with the electric current that CLP had been selling in China since 1979, the project put "China" back into China Light & Power—and put CLP back into China for the first time since it was forced out in 1909. During this time HKE had moved from Wanchai to North Point and then to Lamma Island. CLP built new electricity-generating capacity before there was demand. HKE, with its service area confined to Hong Kong Island and Lamma Island, did not have the same industrial or population growth, and its supply system grew more slowly.

PUTTING CHINA BACK INTO CHINA LIGHT & POWER

Although the specter of China as a political force permeated "1997 and All That," Kadoorie showed little interest in the possibility of China as a market for CLP's current in these essays. Indeed, the 1973 version of the essay, unusually for Kadoorie, is silent on the question of electricity sales to China. Kadoorie may have been mindful that the U.S. embargo against shipments to the PRC had ended only in 1971. The idea that China could develop as a consumer of CLP's electric current and, ultimately, as a location for CLP investments, became an increasingly prominent theme in the two later versions. In the second version he wrote that CLP's and HKE's transmission networks could be integrated into China's national grid: "In the realm of conjecture it is not too difficult to imagine Canton and Hong Kong merging into one continuous area, an important factor in the development of which could well be one or more nuclear power stations, situated in Kwantung [Guangdong] Province, feeding into a national grid that could include the CL&P and HKE distribution systems."[21]

This proved a prescient speculation. At a time when Mao still ruled one of the world's most hermetic and unpredictable countries, Kadoorie imagined that the integration of China and Hong Kong could be bolstered by electricity supply companies. His vision was of an integrated Pearl River Delta region, one that built on the links between Canton and Hong Kong that had existed in the treaty port era to connect the Delta's previously unconnected places. More than four decades later, Kadoorie's vision of regional integration has been enshrined as part of China's national policy in the form of the Greater Bay Area, comprising the Pearl River Delta and surrounding areas. By the time of his third and seemingly final revision seven months later, in January 1976, Kadoorie was even more specific about trying to move into China: "In the light of the present Energy Crisis, and of China's seeming willingness to help in solving Hong Kong's fuel problems, the possibility of building a 1200 MW nuclear power station (two 600 MW units) or a coal-fired station of equal capacity, should be given careful consideration."[22]

Kadoorie's unstated implication, that these proposed nuclear power plants would be built in China, would have been apparent to his readers, given that Hong Kong had recently ruled out the possibility of building a nuclear power plant in the colony. Kadoorie added that this investment in new nuclear or coal-fired power stations would ideally be made in the early 1980s in a joint undertaking involving the Chinese government, the Hong Kong government, and HKE "as participants" along with CLP.[23]

Kadoorie's emphasis on China is understandable, despite the paucity of commercial ties between the PRC and the colony. CLP's first venture had, as noted, ended unhappily when the company was forced to sell its electricity system there in 1909. In both the 1930s and the 1940s the company negotiated to sell current to China; in both cases the negotiations were unsuccessful. In the 1930s the effort foundered over the unwillingness of the potential partner to surrender his opium-importing privileges to a local warlord in exchange for permission to secure the right to import electricity from CLP.

The inauguration of CLP's expanded Hok Un electricity-generating facility in 1940 provided another illustration of the company's ties with and interest in China. A photo of the opening shows three large flags hanging from the ceiling. The central one is British, the other two are the flag of the Republic of China. This was at a time when Japanese troops controlled neighboring Guangdong. Immediately after the war, in his September 1945 letter to Stephens, Kadoorie professed himself optimistic about CLP's prospects in large part because of the possibility of selling electric current to the mainland ("We shall be able to supply electricity to Canton and other points on the Chinese Mainland").

In the late 1940s, after negotiations had reached an advanced stage, Kadoorie professed himself unwilling to pay the bribes needed to sell electricity across the border.[24] After 1949, it soon became impossible to think of selling electricity to China. Simply expanding quickly enough to meet Hong Kong's needs was a challenge for CLP. Moreover, the United Nations embargo on the trade of strategic goods with China, as well as a separate U.S. embargo on virtually any trade with the PRC, would have complicated any such efforts. Indeed, as noted above, Esso received a comfort letter from the U.S. Treasury Department stating that it would not be in violation of U.S. sanctions if it agreed to invest in the joint venture with CLP in Hong Kong.[25]

The PRC's seizure of Kadoorie China assets, ranging from the family's palatial Marble Hall residence to its hotels, could have acted as a further discouragement. Still, the enmity many Chinese and British felt toward the People's Republic is nowhere to be found in Lawrence Kadoorie's public statements. Kadoorie noted that his father, who had died at Marble Hall in 1944 while under detention by the Japanese, was particularly fond of the residence. He went on to say that after the takeover of China by Communists, Marble Hall "has become a Children's Palace" and was "now used by thousands of happy children."[26]

Understanding Kadoorie's past is key to framing a response to the question of how Kadoorie could have foreseen with such accuracy the establishment of a Chinese-ruled, postcolonial Hong Kong with, at least in its initial years, a high degree of autonomy. The answer lies in his ability to draw on the past for inspiration. Born in 1899, Kadoorie looked back to a time of treaty ports and fluid borders, to a world that was, to use today's language, more globalized than that of the 1970s. His father had come to East Asia from Baghdad via Basra and Bombay and worked in various treaty ports, finally living in Shanghai, London, and Hong Kong.[27]

Lawrence was born in Hong Kong and grew up, first in the colony, then in Britain, and later in Shanghai, where he spent much of his early working life. He was acutely conscious of borders, and of the importance of British administration and British law in making Hong Kong a unique enclave. He was a man who moved across those borders with ease. He appreciated the value of borders in enforcing separation and preserving differences, but he was also aware that they were precarious and could be swept away. The Shanghai of his childhood was made up of foreign concessions, each with their own extraterritorial laws. That extraterritoriality was overturned by the Japanese in 1937 and then eradicated after 1949 by the PRC. In Hong Kong, the border had been open for most of the time since the British took possession in 1841. The establishment of the PRC and a dramatic increase in the number of refugees prompted the Hong Kong government to establish a hard border in 1951, largely sealing the colony off from its neighbor.

Given Kadoorie's experience in Shanghai and Hong Kong, it seems plausible that he conceived the idea of what became the Hong Kong Special Administrative Region through his experience of colonies and treaty ports, "neutral" places like Hong Kong and Shanghai where foreign laws could prevail but Chinese and foreigners could mix freely. There was an important difference in Kadoorie's solution, one that set him apart from the ranks of colonial officials, British taipans, and even Chinese business counterparts. They could not conceive of such a world unless it was a colonial enterprise—one composed of British, or at least Western, administration, laws, and leaders.

Where the treaty ports had been dominated by foreign laws and foreign businesses, Kadoorie could conceive of an inversion of that traditional concept, one where Chinese administration would preside but accommodate foreign interests. Kadoorie's was an exercise in historical imagination, one that used knowledge of the past to imagine, and to shape, the future. His vision could have been seen as sycophantic or as an unrealistic, overly optimistic attempt to imagine doing business with a hostile government. Instead, his analysis proved prophetic and of great benefit to CLP.

PROJECT '77

Kadoorie wrote his "1997 and All That" essays with a view to understanding what the end of the British lease on Hong Kong would mean for the territory and particularly for the owners of capital-intensive assets like CLP's electricity generation and transmission system. He circulated the final of the three versions in January 1976. By December 1976, despite some very negative analysis in his essay, he had announced to the board of directors that a new electricity-generating station was needed.

Kadoorie dubbed this Project '77, and in March 1977 he met with Prime Minister Callaghan to advance what became the Castle Peak project. Callaghan, who would be ousted by Thatcher two years later following the public-sector strikes during Britain's Winter of Discontent, proved eager to support British manufacturers by backing Castle Peak. He also expressed interest in Kadoorie's idea of bartering British mining equipment for Chinese coal. At the same time Kadoorie began exploring the possibility of selling electric current to China, an undertaking that bore fruit in 1979.[28]

Yet, to a reader of the final version of "1997 and All That," neither of these undertakings are obvious outcomes. There is a gap in the historical record in 1976 that requires further research in order to help us understand the thinking of Kadoorie and those around him in deciding to invest in the new generating station. The answer may be as simple as Kadoorie extrapolating the expected growth rate of electricity sales and applying the 13.5 percent return on fixed assets that CLP was entitled to under the SoCA to calculate that, even after allowing for high inflation, CLP shareholders could count on both a return of capital and substantial profits before the terminal date of 1997, when Kadoorie expected that all assets in the New Territories would be nationalized. If there were such a simple explanation, then the puzzle remains why he spent so much time over a period of more than two years writing and rewriting "1997 and All That." He may have been determined to forestall any opposition by Esso executives.

Also puzzling is Kadoorie's intention to push ahead with Project '77 and the substantial capital investments that it required, given his conclusion that some sort of government takeover would occur. When Kadoorie teased out the implications for CLP shareholders in "1997 and All That," the situation looked less promising than his self-described optimistic assessment of Hong

Kong suggested. Kadoorie considered the question of whether CLP would be, in his words, a "wasting asset" or a "living concern." Kadoorie came to the conclusion that the company needed to get its capital out by 1997; by that time the generating plants and transmission network were likely to be owned by the government or perhaps a consortium that included China.[29] After 1997, Kadoorie wrote, CLP was more likely to have a management contract than to own the electricity generation and transmission assets in Hong Kong.[30] Selling the company to a consortium that included China "say around 1985 hopefully," or a decade after he was writing "1997 and All That," was a possibility.[31] CLP had to be prepared for the government to appoint one or more directors. Buying assets abroad, in the United States or Canada—Kadoorie specifically mentioned Canada's Churchill Falls, where a large hydroelectric facility was under construction at the time— seemed like a promising way to redeploy shareholder funds.[32] Kadoorie had a strategy to profit from what he expected would be China's pragmatism. He also had an exit plan in the form of a full return of capital before 1997 and alternative locations for future investments.

Kadoorie's downbeat analysis went beyond what the PRC's takeover of Hong Kong would mean for CLP to encompass issues of energy supply and security. The first version of his study was finished in October 1973, the month of the Yom Kippur War and the concurrent OPEC oil embargo. The oil embargo was initiated immediately following the announcement on October 16, 1973 that oil prices would be raised; the embargo remained in place until March 1974. We do not know when the first draft of "1997 and All That" was completed, but it was most likely before the oil embargo was announced, given Kadoorie's silence on the subject. The embargo and the sharp increase in prices were the catalyst for higher inflation and increased concern about energy availability, prompting uncharacteristic worry on Kadoorie's part. In the June 1975 version of his essay he was warning of a Weimar-style collapse:

> In a rapidly changing world with inflation almost out of control and no true base for a stable currency, intelligent anticipation is required now if we are to avoid the dangers of panic and chaos resulting from massive unemployment and labour unrest. . . .
>
> To my mind, having seen the total collapse of paper money in Germany, in Russia, in France, in Italy and in China, inflation is a disease bringing

complete collapse of monetary values and in its wake dictatorships, war and other drastic changes.[33]

Kadoorie in this passage integrates his economic concerns as a businessman with fears of political and social dissolution. Despite this analysis, or perhaps in an effort to forestall a collapse that he otherwise thought would result, he pressed ahead with his largest-ever investment plans. Notwithstanding his fears about the state of the global economy and the threat that CLP would no longer own its Hong Kong generating and transmission assets as a result of some form of nationalization, Kadoorie saw a future for a large new electricity-generating plant. Where sixteen years earlier he had fought government control over the electricity generation and transmission business, he now welcomed the possibility of government ownership as a way of ensuring certainty, which in turn would facilitate the financing and construction of this larger project.

Kadoorie believed that this approach would necessarily mean bigger companies: "World trends have convinced me that the day is rapidly approaching when comparatively small companies can no longer operate as viable entities. It behoves [sic] us therefore to plan with this in mind."[34] Low-cost finance, suggested Esso executive Fred Westphal, could be provided with government-guaranteed bonds. "New requirements for capital after 1980 might be applied by Government in exchange for low-interest bonds."[35]

These suggestive fragments confirm Kadoorie's evolution from a businessman who prided himself on self-reliance to one comfortable partnering with the state. In 1965, Kadoorie had told Esso executives that after 1945, "we received no compensation for war damage. We have had to create order out of chaos by our own efforts. We knew we could not rely on others to rebuild for us and, in consequence, acted for ourselves." The turning point came during the 1959 ESCC hearings and in the five years that followed, when Kadoorie negotiated the joint venture with Esso and the SoCA with the government. By 1976, as he was thinking about what would become the Castle Peak and Daya Bay projects, he needed the government to help him manage the financial risks of an increasingly large and complex system. His view of relations between the state and business were being radically reshaped to reflect both a changing political environment and the vastly larger capital requirements of an age of abundant electricity provided by nuclear-powered and large-scale coal-fired electricity-generating plants.

In her 1982 address at the inauguration of the Castle Peak generating facility, Thatcher made no mention of Hong Kong's return to China in 1997, of which Deng had so abruptly informed her four days earlier. She simply praised the colony as "a unique model of energy and economic driving force in the region." Turning to the power plant itself, Thatcher called the Castle Peak project "an impressive illustration of the blending of British and Hong Kong skill, technology and enterprise" and, as noted, claimed that it presented an "ideal opportunity for the British Government and British industry to work together and to produce a complete package for a project contributing to Hong Kong's continued expansion and development" before lauding "my friend Lord Kadoorie." In a subsequent letter to Kadoorie, Thatcher called the electricity supply station "a most important development for Hong Kong, a success story for British industry, and a magnificent achievement by you and others involved in bringing this great project to its present stage."[36]

Thatcher's 389-word speech looks, on first reading, like little more than a series of platitudes. Its importance ranking in the Thatcher Archive, where it is filed under the themes of "Industry, Energy, Foreign policy (Asia)," is characterized as "Minor." But the opening of Castle Peak had a greater significance than its relegation to an archival footnote would indicate. For one, the opening of the power station reconfigured political and business relations between Hong Kong and Britain during a period of intense political anxiety for Hong Kong, ensuring continued high-level British involvement in the colony. Second, Castle Peak led the way to a dramatically increased role for Hong Kong in China. Together with CLP's other generating facilities, the power plant acted as a bridge to China, providing electricity and paving the way to what three years later would be CLP's investment in the PRC's first commercial nuclear reactor. CLP's capital, representing a 25 percent ownership stake in the Daya Bay nuclear plant, would constitute a significant portion of what was at the time China's largest direct foreign investment.

Kadoorie's "1997 and All That" was a serious and time-consuming study. Kadoorie concluded that CLP would likely have to relinquish ownership of its generating plants and transmission network by 1997; yet the allusion to a lighthearted history book that told a story of greatness despite takeover by a foreign power suggests that he was convinced the company could somehow not merely survive the vicissitudes of political transition but thrive, just as Britain itself had. His essays provide an unusual and previously unknown perspective onto Hong Kong's relations with China and electricity's key role in the economic integration of China and Hong Kong. At the end of "1997 and All That," Kadoorie wrote that he hoped to see directors after the expected takeover of Hong Kong by the PRC. "These private notes are prepared in anticipation of our Annual General Meeting in 1998—a meeting I hope my colleagues will make every effort to attend."[37]

Chapter Eight

SING THE CITY ELECTRIC

Hong Kong underwent profound change between the end of World War II and British Prime Minister Margaret Thatcher's pioneering visit to Beijing in September 1982. From an economically peripheral, war-ravaged city of 500,000 people in 1945, the colony by 1982 had a population of more than five million and was a significant global manufacturing and financial node. The colony's electricity supply system had a little-examined importance in making Hong Kong the quintessentially modern Asian hub that it became during those four decades. Most histories of Hong Kong have relegated technology to the background or treated it as a reflection of a generalized process of modernization. Pushing technology to the foreground allows us to better understand the transformative impact of a macroinvention such as electricity and to see why the colony emerged as a place of global significance.

This book has looked at three central questions. What role did electricity play in Hong Kong's economic development and its internal social and political transformation? How, in turn, did the colony's changing economic, political, and social circumstances—from the ruins of war in 1945 to the start of negotiations about reversion to Chinese sovereignty in 1982—influence the development of this essential utility system? And finally, what can the history of CLP's evolution tell us about the interplay between the state and private businesses in the four decades after the end of World War II, both in the colony and more broadly?

CHINA LIGHT & POWER IN HONG KONG'S DEVELOPMENT

Hong Kong was one of the world's fastest-growing economies during these decades, expanding about 7.8 percent annually from 1948 to 1972.[1] It acquired a head start on its East Asian neighbors in economic growth in large part because of the relocation to Hong Kong of Shanghai textile entrepreneurs, whom Lawrence Kadoorie wooed with promises of ample supplies of electricity and who in turn were responsible for much of the export-oriented light manufacturing that was so important for economic growth in the forty years after World War II. Hong Kong's rapid economic growth transformed the British-Chinese colonial project, making Hong Kong both Britain's competitor and its customer.

More broadly, Hong Kong's fast-paced economic expansion helped reshape the West's image of Asia. Hong Kong showed that Asia beyond Japan could become an economically vibrant region that would play an important role in the global economy. Its low-priced manufacturing exports and the attempt by British competitors to restrict them proved a harbinger of a new era of East Asian, and of Chinese, economic strength and the tensions that would result from an economically more assertive Asia.

Kadoorie's vision of an electrified Hong Kong, and the bountiful electricity supply system whose creation he oversaw, was a necessary condition for Hong Kong's transformation from a trading entrepôt to a significant export-oriented manufacturing base and global financial hub. An ample supply of electricity, a fundamental technology that has unusually wide-ranging applications and implications, was a prerequisite for the high economic growth that became a defining feature of Hong Kong's image. Kadoorie's understanding of the possibilities electricity promised for the colony and his determination to provide enough electricity to enable manufacturing-led growth and employment and thus forestall social unrest were important strands in the articulation of a new vision for Hong Kong. So, too, was his understanding that electricity-generating stations could be important pieces in the geopolitical contest that was being played out in and around Hong Kong between Britain, China, and the United States.

What occurred in Hong Kong in the four decades after 1945 was more than a dramatic material transformation. It also revolutionized how Hong Kong people thought about the colony, how the world perceived Hong Kong, and how Hong Kong could act as a sort of airlock between Communist

China and the West. In anticipating Hong Kong's need for electricity and the dramatic economic changes instituted in China toward the end of the period under review, Kadoorie both accelerated and amplified those changes. He created a future that reflected his singular vision by building CLP's electricity-generating stations in the colony and at Daya Bay, just across the border from Hong Kong in mainland China.

CLP's electricity enabled Hong Kong's modernization. Its electric current provided the light and power that generated one of the world's fastest-growing economies. Electricity also made possible other facets of a modern city, creating the economic and social conditions that prompted a radical transformation of society. CLP provided electric current for the housing estates and hospitals and schools that were built to serve the city's expanding population, for the high-rise buildings accessed by elevators, for an extensive electric-powered underground transit system (the Mass Transit Railway or MTR), and for brightly lit shops and restaurants. All this combined to produce a new sense of the colony and its place in the world, one whose bright lights blazoned capitalist abundance.

CLP was one of Hong Kong's largest and most sophisticated technological undertakings and at the same time the product of specific business choices made by Kadoorie and CLP directors and managers. Aside from the more than hundredfold increase in electricity-generating capacity, and the many social and economic ramifications that resulted from this supply, Kadoorie used the electricity supply network for strategic political advantage. Notable was the relationship between CLP's electricity network and the colony's political, business, and economic development during a time of rapid economic growth, particularly his insertion of CLP into Sino-British relations.

A powerful demonstration of Kadoorie's belief in the geopolitical potential of the CLP electricity system was his promise of additional investments to secure an important role for Hong Kong in the run-up to and aftermath of Hong Kong's reversion to China in 1997. The result was the Castle Peak coal-fired electricity plant, as well as a gas-fired plant that opened in 1989 at Black Point in Hong Kong's New Territories and the Daya Bay nuclear-generated electricity complex in China's Guangdong province, which started production of electricity in 1993. In envisioning these projects, Kadoorie looked forward to a new post-1997 world in which Hong Kong would be part of China. At the same time, he drew on patterns that had been imprinted in his early life, images of treaty ports and other enclaves that were within China but maintained a separate foreign identity and were useful to China precisely because of their otherness.

Kadoorie's longstanding goal was to bring CLP's electric current, and CLP as a company, back to China. This he did in 1979, the first time the company had sold current to or in China since being forced out in 1909. Kadoorie had long been confident that he could accomplish this goal because of his belief that the Chinese government would recognize the unique role that colonial Hong Kong played in furthering China's development. He believed that if Hong Kong could show that it was useful to China, even a Communist government would protect the colony's status as an enclave where secure property rights and efficient government administration would facilitate investment.

He also worked to keep Britain fully engaged in the colony in the run-up to the 1997 handover as a way of ensuring the longer-term protection of Hong Kong and of his assets there. He negotiated financial arrangements making CLP's and Esso's investment in the new Castle Peak generating facility profitable even though he believed that nationalization or state management of CLP's electricity supply network would occur no later than 1997. Thatcher's ceremonial opening of the Castle Peak plant in September 1982, just four days after Deng Xiaoping formally told the British government that China would take full control of Hong Kong, underscored the symbiotic relationship between electric and political power.

Kadoorie's final project, the Daya Bay electricity-generating station, China's first civilian nuclear reactor and the country's largest-ever foreign investment at the time it was built, was already under discussion at the time Thatcher went to Beijing in 1982; it was a point of discussion then and on her subsequent trip in 1984.[2]

The Daya Bay nuclear reactor complex was a highlight of, and in some senses exemplified, the early years of China's economic opening and reform; Daya Bay's operations in turn provided the electricity to enable accelerated economic growth in the southern province of Guangdong, much of it used by Hong Kong-invested factories. French and British corporate and government archives would yield additional information. So would Hong Kong sources, including media, given the controversy over building a nuclear power plant so close to the colony. The construction occurred in the late 1980s and early 1990s, a period that was overshadowed by the 1989

Tiananmen Square killings and was a time of particular anxiety over the return of Hong Kong to Chinese sovereignty. This confluence of events adds to the sociopolitical importance of the Daya Bay project.

Kadoorie had long seen electricity as a means of promoting historical progress, as was evident in his 1940 speech at the Hok Un station and subsequent speeches and articles. After 1945, he articulated more fully his vision of electricity as a tool of economic growth and a means of ensuring social stability. He expanded CLP's supply network and saw to it that industrialists obtained adequate electricity to establish factories that would provide jobs for workers. CLP's electricity pricing favored factories to such an extent that there were times when Kowloon factory owners sometimes did not even pay for the fuel cost, let alone the operating or capital costs, of the electricity they consumed.

Kadoorie chose to adopt aggressive and opaque pricing policies to finance the high growth in electricity supply that he correctly anticipated would be needed to realize his vision of a more socially stable and implicitly depoliticized Hong Kong. He saw opportunities for profit, to be sure, but he also saw the necessity of providing the electricity needed to power Hong Kong's peaceful industrial revolution.

The example of CLP shows how technological diffusion occurred in a colonial post-1945 context. CLP was embedded within a far-reaching system that had deep local penetration as well as a global reach. It included equipment suppliers and financiers in Britain, New York City–based oil giant Esso, as well as coal suppliers from around the world. Kadoorie and his managers at CLP assembled generating and transmission equipment, fossil fuels, managerial expertise, and capital from abroad along with local Hong Kong elements to produce profitable and high-speed growth for the company and for the colony. Hong Kong had no indigenous research capability, yet CLP built a sophisticated and large-scale electricity system by assembling parts from a global network built by Kadoorie and others at CLP.

The most transformative global linkage was with Esso, one of the world's largest petroleum companies. When Esso signed a joint venture to build a new electricity supply station with CLP in 1964, the U.S.-based multinational was undergoing a long process of reassembling itself following the breakup of the Standard Oil trust in 1911. Esso had ended its joint venture marketing agreement with Mobil and was looking for new customers in the region. As a large and demanding company that was used to imposing its systems on partners, Esso forced CLP staff and Kadoorie himself to be more explicit and systematic in their management procedures. These included an extensive use of computer processing and detailed and explicit personnel policies; the latter were procedures necessary to run the larger and more complex company that CLP would become. This investment facilitated a significant expansion in Hong Kong's electricity-generating capacity.

Esso's investment in and board representation on the joint venture electricity generation operation inevitably diminished Lawrence Kadoorie's role at CLP. The resulting professionalization of management ultimately eased the transition to a post–Lawrence Kadoorie era. Esso's capital and the predictability that came with the SoCA allowed investment on a far larger scale than had been possible.

A principal attribute of the colony of Hong Kong was its natural harbor. For that harbor to be useful for merchants, infrastructure of various sorts had to be built—piers, roads, a water supply network, and, later, a telegraph system and electricity network. From the colony's earliest days, the need for high-quality infrastructure to serve trade preoccupied colonial administrators and businessmen alike. Without good infrastructure Hong Kong would not prosper as the emporium for China trade.

CLP's electricity supplies ensured that Hong Kong had the ability to function, first as an entrepôt and later as a globally efficient exporter. That CLP's electricity supply network was built and operated by a private company, albeit one that eventually had to accommodate to government control, showed Hong Kong's hybrid nature.[3] In Hong Kong's case, this meshing had significant political and economic elements as well as cultural ones. The hybridity, a legacy from more than 150 years as a crown colony, underpinned the establishment of the Hong Kong Special Administrative Region in 1997.

Hong Kong did not take a neoliberal turn in the late twentieth century, for it had no need to: it had deviated relatively little, in a global context, from its nineteenth century hands-off government policies. Yet, as this study of CLP has shown, the nature of the business-government relationship deepened and became more structured in the post-1945 period.

Hong Kong's elites mobilized technology to affirm the colony's modern identity. Concern with and pride in the quality of Hong Kong's engineering and building was evident from the nineteenth century, notably with the Praya reclamation of the 1890s. CLP, beginning with the Kadoories' ascendancy in the 1930s, "bought of the best," priding itself on building an ever-larger electricity-supply network in advance of demand. The colony's business and political elite's concern with high-quality technology and infrastructure intensified in the post-1945 period, in part to forestall social unrest and stave off demands for more sweeping political reforms. Electricity offered a path that promised economic and technological development but did not demand far-reaching political change. It both upended the old social and economic order even as it allowed the political structure of elite colonial rule to remain largely unchanged.

The sextupling of the population in the fifteen years after the war necessitated new arrangements. A populace that had been toughened by experiences of war and displacement and influenced by greater state involvement in economies throughout the world, demanded more from government than the prewar colonial order could offer. Social reformers' demands for change, given further impetus by the fear of fires and diseases, could not be ignored. Hong Kong's development of public housing started before the 1953 Shek Kip Mei fire, but that event catalyzed an acceleration in the provision of housing and other public services. The building in the 1950s of Queen Elizabeth Hospital, the Commonwealth's largest, came against a backdrop of concern about high rates of diseases such as tuberculosis and typhoid.

CLP's network of generating stations and high-voltage transmission pylons, its substations and meter readers, combined to play a performative role, one that literally displayed the infrastructure necessary to make a modern life possible. CLP and HKE facilities played spectacular roles during royal coronations and other celebrations. Day in and day out, the company's system and its technologies both exemplified urban, industrial modernity and provided the current that made other manifestations of modernity possible. The idea of the colony's infrastructure technology, including engineering, construction, and electricity-generating technology, embedded itself as a core Hong Kong value during the 1950s and 1960s. Ample, high-quality infrastructure during this time became part of the colony's normative self-image. One example is the expansion of Kai Tak Airport in the late 1950s and early 1960s, a response to the desire to increase trade and facilitate exports. As with public housing and Queen Elizabeth

Hospital, Kai Tak's expansion was paid for entirely with government funds. Government financial support, in the form of both a grant and concessionary loans, made possible the construction in the 1960s of Ocean Terminal, another important node in Hong Kong's network of transport linkages. Modern factories set up by industrialists who had fled Shanghai constituted another source of growth.

Most of Hong Kong's elites, Chinese and non-Chinese alike, consciously distanced themselves from political and social changes taking place in Britain and elsewhere, such as fuller rights for women and a broader electoral franchise. Political reforms proposed by Governor Mark Young were not approved, with key opposition coming from the Chinese business elite, but pressures for change remained. Reforms such as compulsory state-funded education would largely have to wait for the 1970s and the era of Governor Murray MacLehose.

MODERN HONG KONG AND THE CONTROL OF CHINA LIGHT & POWER

Although electricity was one of the preconditions for the postwar economic liftoff, that very economic success had, by the late 1950s, in turn imposed new social and political pressures on the electricity supply companies. Changing social, political, and economic expectations in Hong Kong collided with the financial needs of CLP's rapidly expanding electricity supply system to produce a moment of intense political conflict that culminated in the 1959 ESCC hearings and the debate that followed over the proposed nationalization and merger of the two electricity supply companies. A new shared social, political, and economic sense of electricity as a privately produced public good was shaped in the crucible of the ESCC hearings and their aftermath.

The 1959 ESCC hearings and the ensuing political debate and bargaining marked a significant yet largely unnoticed moment in Hong Kong's sociopolitical and economic history. The recommendation by an elite government-appointed commission to merge and nationalize the two electricity supply companies, and the decision by officials including Financial Secretary John Cowperthwaite to use dividend restrictions as a weapon to control CLP, reflects the changing political-economic framework of the postwar decades. The hearings and their aftermath also illustrate the relationship between powerful individuals such as Lawrence Kadoorie and larger systems, including the colonial state's administrative apparatus and CLP's electricity system.

Kadoorie's vision was elitist and, in many senses, paternalistic. He was determined to show Hong Kong's usefulness to both China and Britain. Showcasing the colony as a producer and a consumer, the Peninsula Hotel placed the largest-ever order for Rolls-Royce automobiles, and the construction of CLP's Castle Peak generating station constituted the largest-ever export order for British power manufacturers. Electricity, in Kadoorie's view, would also minimize the danger of social unrest by providing factory employment. His success in realizing his vision led to a collision with the social and political aspirations of the richer, more complex society that CLP's and HKE's electricity had helped create. This vision had to be modified, as he accepted the need for government restrictions. Indeed, he ultimately appreciated the way in which the restrictions brought predictability and thus more financial certainty, allowing him to borrow funds as never before. He also revised his thinking about the autonomy of private business and his earlier belief that government should take only the most limited role in the economy; he came to welcome government involvement in sharing the risk of large-scale electricity-supply systems.

The history of CLP demonstrates key facets of the colony's transformation. Hong Kong's electricity supply arrangements influenced the development of a larger role for the state in housing, health, and education; the emergence of a more robust civil society; and electricity's role in powering the economic growth that underlay these changes. Lawrence Kadoorie—in his role as CLP's chairman—shaped this narrative, even as he changed his views over time. Other voices joined the unprecedented public debate on electricity supply arrangements; and the politically negotiated outcome, which saw Esso join forces as CLPs partner, was quite different from what Kadoorie had initially proposed. In this controversy we can see the articulation, contestation, and development of a new vision for the colony.

The ESCC hearings elevated the question of electricity supply to that of a political project and, through that exercise, nurtured the emergence of civil society in Hong Kong. The war effort of World War II led to a larger role for the state and a questioning of prewar orthodoxies in Britain, in Hong Kong, and elsewhere. In Hong Kong, the tremendous population pressures, as well as the proxy war being played out between the KMT and

CCP, made this rethinking more acute and far-reaching. After 1945, social and political issues in Hong Kong could no longer be repressed as effectively as before 1941. The public housing program and the expansion of medical services reflected a reluctant acceptance of government responsibility for social welfare issues. A salaries tax provided funds to expand the size of government. By focusing on a large state-like entity, CLP, as an object of political and social control, the government reduced the pressure it was under to do more. In Hong Kong as in Quebec, imposing political controls on electricity supply companies provided "a sort of safety valve" at a time of "modern communication, more general education, and more freedom."[4] In Hong Kong, the charged moment of the ESCC hearings and their aftermath helped catalyze change and at the same time crystallize a new sense of expectation of what it meant to be well governed.[5]

Choices were made to initially favor industrial electricity users, a decision later recalibrated because of the social and political pressure exerted by the ESCC hearings. Providing electricity for domestic consumers was an afterthought, and CLP did little to promote the use of electricity by individual users. In comparison with a global leader in urban electrification, such as Samuel Insull's Chicago Edison, or even with the colony's own HKE, CLP was little interested in stimulating consumer demand for electricity. Like many electricity supply companies, it resisted serving rural consumers. Notwithstanding Kadoorie's promises to supply electricity whenever and wherever it was wanted, and in whatever quantity it was desired, New Territories villagers were subjected to ongoing discrimination in the provision and pricing of electricity supply. In this discrimination against rural customers, CLP's policy had many global precedents. As in many other places, only after the application of government pressure did CLP extend electricity supply to all rural areas.

CHINA LIGHT & POWER AND STATE-BUSINESS RELATIONS IN HONG KONG

A close look at CLP's history shows such an extensive intertwining of government and business interests as to raise the need to challenge oft-repeated claims that Hong Kong, a colonial free port, exemplified a laissez-faire economy. The experience of World War II and the mobilization of entire societies established a pattern of state involvement in the economy and society that was extended after the war. Hong Kong was unusual in that its government was less interventionist than that of many newly independent colonies, particularly where tradeable goods were concerned.

Measured by the standards of the colony's prewar period, the Hong Kong government after 1945 was deeply involved in society and the economy, as epitomized by the introduction of a salaries and profits tax in 1947. If Hong Kong was a relative exception, it was one that nonetheless proved a rule: modern economies, with their technology-rich electricity and other complex systems and more demanding citizens, required a new level of government control. Claims to a hazy laissez-faire Eden on closer inspection turn out to be a product of wishful thinking by distant observers.

The intricate ties between Kadoorie and political leaders and administrators show the development of this hybrid state-business model. The existence of close relationships between the colonial business and political elite is well known, but this phenomenon has typically been expressed as a general assertion rather than supported with detailed evidence. Archival documentation reveals the ongoing political, business, and social links between Kadoorie, CLP, and the Hong Kong government as well as between Kadoorie and the most senior politicians in Tory and Labour governments. This was evident in the dozens of documented meetings and phone calls between Kadoorie and Financial Secretary John Cowperthwaite during the nearly four years it took to negotiate the SoCA as well as in Kadoorie's meetings with Prime Ministers Callaghan and Thatcher and a number of cabinet ministers.

A detailed examination of the ESCC hearings and their aftermath reveal a granular, nuanced version of Hong Kong's political-economic policies. The historical record also demonstrates the idea of continuity, for the idea of state control of the electricity supply system had antecedents that stretch back to the 1920s. Military control and operation of the electricity supply system in 1945–46 provided a template for state involvement in electricity supply arrangements. An attempt in 1948 to control tariffs and dividends was accompanied by a warning that there was no hurry, as "the legislation contemplated is a trend of the times." The colonial government played a long game regarding control of the electricity supply system. Yet Kadoorie played a longer game still. The CLP system outlasted him. It remains privately managed and owned, with the Kadoories as controlling shareholders, to this day. The SoCA remains in place in Hong Kong, its essentials largely unchanged from what was introduced in 1964. Outside of Hong Kong, CLP is a significant investor throughout Asia and Australia and constitutes one of the region's largest investor-owned electricity producers.

The history of CLP shows the importance of technology, and more specifically the electricity supply system, in Hong Kong's colonial history. Electricity had a central role in the political, economic, and sociocultural development of colonial Hong Kong in the four decades after the end of World War II. And yet it was one person's vision—that of Lawrence Kadoorie—that both reflected and also helped shape a collective belief in an ideal of electric-powered economic and social progress, even as parts of that vision were rejected and then transformed and in turn transformed Kadoorie's own thinking. The story of Kadoorie and CLP speaks to the way power—both political power and technological power in the form of electricity—contributed to the shaping of a new and radically altered sense of colonial Hong Kong and its future up to 1997 and beyond. In post-1945 Hong Kong, electricity was a key element that allowed for a rearticulation of what the colony represented in an increasingly postcolonial world.

Electricity is an element that underpins modernity and modern life. Electricity's many manifestations, from transmission pylons to illuminated signs to elevators, both symbolize and tangibly demonstrate what it is to be modern. Electricity represented the triumph of technology and bolstered a sense of historical progress. Traditional ways of doing things in government, in society, and in business were questioned and often transformed. The ESCC hearings and the debate that surrounded and followed them; the ceremonies opening new or expanded electricity supply stations; CLP commemorations of royal events; and, of course, the provision of electricity itself—all of these played their part in the territory's transformation. As an indispensable part of a larger social and political undertaking, electricity helped to establish the modern world, redefining the urban experience in Hong Kong and beyond.

ACKNOWLEDGMENTS

This study grew out of my book *The Greening of Asia*, a study of the changing energy landscape in Asia, whose penultimate chapter featured China Light & Power. CLP interested me because of its 2007 decision—at the time unique among its electricity supply company peers—to cut its carbon intensity 75 percent by midcentury. I happened to be present when CLP's chief executive, Andrew Brandler, made the announcement. Several years later, Andrew helped crystallize my thinking for what became *The Greening of Asia*. Subsequently, Sir Michael Kadoorie; Andrew; Andrew's successor as CEO, Richard Lancaster; and many others at CLP talked candidly to me about how the company was trying to achieve that goal, sharing stories of stumbles as well as triumphs.

When I wanted a different sort of midlife challenge, Robert Peckham at the University of Hong Kong's History Department generously offered it by agreeing to support me as I studied what the development of Hong Kong's electricity system meant for colonial modernization. Robert is patient yet provocative, a meticulous reader and editor, one who overflows with new ideas, new ways of looking at a problem, and with new books to read. Others at the university who have supported my work include John Carroll, Frank Dikötter, and Peter Cunich. John provided incisive comments on the manuscript.

ACKNOWLEDGMENTS

The Centre for the Humanities and Medicine and the Hong Kong Institute for the Humanities and Social Sciences both hosted talks that I gave on my research, as did the History Department. My thanks go to Ria Sinha for organizing the Centre's two seminars where I presented my work and to Angela Leung and her colleagues at the Institute for making my talk possible, as well as to all of those who took part.

For *Greening of Asia*, I looked at CLP at the turn of the twenty-first century. Little did I imagine how much more time I would spend examining the company's nineteenth-century antecedents and its growth through the twentieth century. I hadn't intended to return to CLP after I finished research for *The Greening of Asia*. As Robert and I talked over a period of months in 2013 and 2014, however, it seemed natural to look more closely at what now is one of the largest investor-owned electricity supply companies in Asia as a way of better understanding the development of Hong Kong, particularly in the critical decades after 1945. Electricity is a useful way to frame economic, political, and social change, one that is particularly appropriate for looking at fast-growing economies such as Hong Kong's. Electricity was an indispensable prerequisite for Hong Kong's economic development; the colony's rapid growth in turn looped back to shape the system that CLP was building. Electricity and society were intertwined in a dance in which, as William Butler Yeats put it more poetically, it is difficult to separate the dance from the dancer.

Sir Michael, Richard, and Andrew offered their support to this project as well, ensuring that I had full cooperation from others in the company and, above all, those at its archive. I was fortunate in that Sir Michael in 2007 had established the Hong Kong Heritage Project (HKHP) as a repository for histories of businesses linked to the Kadoorie family. Coincidentally, and to my great benefit, CLP papers from the company as well as from the office of Sir Elly Kadoorie & Sons were made available to scholars shortly before I started my research. The staff at the HKHP have been unfailingly helpful and have accommodated my requests for material that is generally not open to researchers as well as in finding a great deal of material. My thanks go to HKHP Executive Director Fanny Iu, Amelia Allsop, Melanie Li, and Sing-ping Lee. No one at CLP or HKHP read this book in manuscript form or ever tried to influence my work in any form. I have had nothing but support from CLP and HKHP staff as they took the time to try to help me better understand the company. The photos are courtesy of HKHP.

ACKNOWLEDGMENTS

Hong Kong is unique among major global cities in having not one but two electricity supply companies that are each more than a century old, each serving geographically distinct areas, thus affording an unusual comparative perspective. At Hong Kong Electric, Frank Sixt graciously made an introduction to its managing director, C. T. Wan, who, along with longtime executive S. S. Yuen and Group legal counsel and company secretary Alex Ng, generously shared company history. Howard Elias, a board member of the Jewish Historical Society of Hong Kong, was a continuing source of encouragement and a rich source of information on Hong Kong's Jewish community. The research librarians at the University of Hong Kong were unfailingly generous in helping me find material from among the university's impressive holdings. I also want to acknowledge help from the staff at the Hong Kong Central Library, where I accessed the Carl T. Smith collection of cards on long-term loan from the Royal Asiatic Society, as well as Hong Kong's Public Records Office (HK PRO). Vaudine England repeatedly and promptly helped with various queries and navigational advice. Reuters archivist David Cutler helped trace the company's early activities in Hong Kong, ones that coincided with the arrival of the telegraph. Allan Pang provided invaluable help by photographing documents at the National Archives in Kew relating to Elly Kadoorie's 1927 naturalization. Dick Yeung drew the maps. Thanks to Warwick Anderson for his quick and positive response to my request to publish several lines from "Unit" for my epigraph. I also benefited from the insights and camaraderie of Jonathan Kaufman, author of *The Last Kings of Shanghai*, a history of the Kadoories and Sassoons.

I have had the fortune to have friends who are also close, careful, generous, and supportive readers in Jill Baker and Pamela Mensch (who has been indefatigable and attentive in her repeated readings of the manuscript), as well as the support of my wife Melissa and son Ted, both of whom read large parts of this manuscript, and my daughter Anya. My final thanks are reserved for two people who came first, Tom Laqueur and Peter Dale Scott, professors whom I met during my undergraduate years at UC Berkeley. They helped set me on an intellectual voyage that continues four decades later. Indeed, it was Tom's mention, at the Musical Offering Café in Berkeley, of an eighty-year-old student of his that catalyzed my long-held desire to do a more academically rigorous project. Years earlier, Tom and Peter had helped define my intellectual interests and, more important, showed me how to take scholarship seriously but not slavishly, to be engaged but not smothered.

At Columbia's Center on Global Energy Policy, Inaugural Fellow David Sandalow and Founding Director Jason Bordoff provided invaluable support. I owe them a deep debt of gratitude for nurturing this book. At Columbia University Press, editor Caelyn Cobb, associate editor Monique Laban, production editor Marisa Lastres, and indexer Alex Trotter made the production process mercifully smooth. My deepest appreciation to Gregory McNamee, a meticulous copyeditor who went above and beyond to help me locate a final handful of stubbornly elusive references. My thanks to all of them for their patience and grace in bringing this project to fruition at a time when circumstances prompted me to make an unplanned and disruptive move from Hong Kong. Special thanks to Carroll Bogert for being with me for the final stretch.

Whatever strengths this study possesses owes much to all of those cited here. I hope that this narrative of how one Hong Kong company journeyed from its modest beginnings to a firm of global scale and importance might better inform the social, political, and economic choices we must make about our energy and environmental futures as we reconfigure our energy systems for a low or even postcarbon era.

NOTES

1. PRIVATE LIGHT AND COLONIAL POWER

1. Margaret Thatcher Archive (hereafter TA), September 24, 1982; E. F. Vogel, *Deng Xiaoping and the Transformation of China* (Cambridge, MA: Harvard University Press, 2011), 490–97.
2. TA, September 28, 1982, "Castle Peak A"; China Light & Power (hereafter CLP), *Golden Memories of Tsing Yi Power Station* (Hong Kong: CLP, 1999), 8.
3. China Light & Power was founded in 1900 as the China Light & Power Syndicate Ltd; from 1901 to 1918 it was known as the China Light & Power Co. Ltd; from 1918 to 1935 as the China Light & Power Company (1918) Ltd.; and henceforth for the period covered here as the China Light & Power Company Ltd. I generally refer to the company as CLP throughout, except in quotations or when there is a need to differentiate a particular corporate form. Furthermore, references in this work to Hong Kong include the colony generally, whose size the Convention of Peking (Beijing) extended in 1860, adding Kowloon, along with a lease in 1898 that added the New Territories as part of the Crown Colony for ninety-nine years.
4. J. Woronoff, *Hong Kong: Capitalist Paradise* (Hong Kong: Heinemann, 1980), 2.
5. L. Winner, "Do Artifacts Have Politics," *Daedalus* 109, no. 1 (1980): 131.
6. S. Jasanoff, "Future Imperfect: Science, Technology, and the Imaginations of Modernity," in *Dreamscapes of Modernity: Sociotechnical Imaginaries and the Fabrication of Power*, ed. S. Jasanoff and S. H. Kim (Chicago: University of Chicago Press, 2015), 4.
7. J. Mokyr, *The Lever of Riches: Technological Creativity and Economic Progress* (New York: Oxford University Press, 1992), 12–14, 295.
8. D. E. Nye, *How Does One Do the History of Technology?* (Detroit MI: The Society for the History of Technology, 2014), 4–5; R. J. Gordon, *The Rise and Fall of American Growth* (Princeton, NJ: Princeton University Press, 2016), 555–57.

1. PRIVATE LIGHT AND COLONIAL POWER

9. L. Kadoorie, "General Policy (A Brief Survey)" (unpublished manuscript, 1938).
10. There has been little scholarship on the history of coal in colonial Asia, with the notable exception of S. X. Wu on coal in late-Qing China: *Empires of Coal: Fueling China's Entry Into the Modern World Order, 1860–1920* (Stanford, CA: Stanford University Press, 2015). This paucity of research is a notable gap given coal's importance to both imperial and colonial economies in the nineteenth and twentieth centuries. To take one example, French policy and ambitions to expand in Indochina and China, which included infrastructure projects such as the building of the Yunnan railway, were bound up with plans for resource extraction. Peter Liberman has analyzed the Japanese imperial extraction of resources: *Does Conquest Pay? The Exploitation of Occupied Industrial Societies* (Princeton, NJ: Princeton University Press, 1996), 99–119.
11. Colonial Office (hereafter CO), 129/150, no. 82, June 10, 1871, records a telegram of June 8 from the Secretary of State on the completion of the telegraph cable on June 9 ("Historical and Statistical Abstract of the Colony, 1841–1930 [20]"), in Hong Kong Government (HKG) *Administrative Reports*, vol. 4, 1920–1930). June 13, 1871, marked the first recorded telegram sent from the Hong Kong governor in the CO/ 129 series. See also A. Coates, *Quick Tidings of Hong Kong* (Hong Kong: Oxford University Press, 1990), 40, and D. Read, *The Power of News: The History of Reuters* (Oxford: Oxford University Press, 1999), 63. G. B. Endacott writes that the first telegraph communication with Europe occurred in 1870 with a submarine cable to Shanghai, where it joined the Danish Trans-Siberian landline. Inside Hong Kong, Endacott notes that Jardine, Matheson & Co. had connected its Hong Kong offices with a telegraph system in 1866 and that police stations soon thereafter linked to one another. Endacott, *A History of Hong Kong* (Hong Kong: Oxford University Press, 1973), 158.
12. E. Baark, *Lightning Wires: The Telegraph and China's Technological Modernization, 1860–1890* (Westport, CT: Greenwood Press, 1997).
13. Hong Kong Heritage Project (hereafter HKHP), SEK-04-029, C.2-0-1/22. For HKHP material, where available I have included the document number, in this case 22, within a file. Where individual documents are not numbered, I have endeavored to provide other information, such as the date or the title, or an adjacent file number.
14. HKHP, SEK-04-029, C.2-0-1/12, 32, 46, 61.
15. HKHP, SEK-04-029, C.2-0-1/1, 2.
16. *London Gazette*, September 25, 1981.
17. HKE claims that it is the world's oldest electricity supply company operating under unbroken management (Alex Ng, personal communication). CLP is ten years younger and also has unbroken management.
18. HKE was self-consciously a local company in contrast to the Gas Company, which was founded and run from London from 1861 until 1954, much to the unhappiness of some members of the Hong Kong business elite. R. Hutcheon, *The Blue Flame: 125 Years of Towngas in Hong Kong* (Hong Kong: Hong Kong & China Gas, 1987), x, 78.
19. Milton Friedman, "Hong Kong Wrong," *Wall Street Journal*, October 6, 2006.
20. HKHP, C-2-U.1/ 97, Appendix D; CLP, Annual Report 1956: 4–5.

1. PRIVATE LIGHT AND COLONIAL POWER

21. J. E. Strickland, *Southern District Officer Reports: Islands and Villages in Rural Hong Kong, 1910–60* (Hong Kong: Hong Kong University Press, 2010), 167, 242. In one example, the Peng Chau Rural Committee provided electricity at $1 per kilowatt hour (kWh) and free street lighting, but the government pressed CLP to take over the local electricity supply (27, 166).
22. T. Hughes, *Networks of Power: Electrification in Western Society, 1880–1930* (Baltimore, MD: Johns Hopkins University Press, 1983), 460.
23. Hughes, *Networks of Power*; J. Lambert, *The Power Brokers: The Struggle to Shape and Control the Electric Power Industry* (Cambridge, MA: MIT Press, 2015); R. C. Tobey, *Technology as Freedom: The New Deal and the Electrical Modernization of the American Home* (Berkeley: University of California Press, 1996).
24. Y. Nakano, *Where There Are Asians, There Are Rice Cookers: How "National" Went Global via Hong Kong* (Hong Kong: Hong Kong University Press, 2009).
25. CLP, Annual Report 1984: "Ten Year Summary: Statistical Highlights."
26. CLP, Annual Report 1982.
27. E. F. Szczepanik, *The Economic Growth of Hong Kong* (London: Oxford University Press, 1958), 107; S. Tsang, *A Modern History of Hong Kong* (Hong Kong: Hong Kong University Press, 2004), 161–79, esp. 163; M. Kadoorie, author interview, December 7, 2016.
28. M. Kadoorie, author interview, December 7, 2016.
29. A. W. Grantham, *Via Ports, from Hong Kong to Hong Kong* (Hong Kong: Hong Kong University Press, 1965), 104.
30. D. Yang, *Technology of Empire: Telecommunications and Japanese Expansion in Asia, 1883–1945* (Cambridge, MA: Harvard University Asia Center, 2008); F. Dikötter, *Mao's Great Famine: The History of China's Most Devastating Catastrophe, 1958–1962* (New York: Bloomsbury Press, 2010).
31. Lynn Pan and Trea Wiltshire, *Saturday's Child: Hong Kong in the Sixties* (Hong Kong: Form Asia, 1993), 5.
32. HK PRO, *This Is Hong Kong*, film 0029-E055 (1961). On the conscious crafting of the Hong Kong image in this period, see J. Fellows, "Crafting Hong Kong's Image Overseas: 'Commercial Public Relations' in Hong Kong, 1962–1966" (unpublished paper, University of Hong Kong Spring History Symposium, 2016), 5–10.
33. World Bank, *The East Asian Miracle: Economic Growth and Public Policy* (Oxford: Oxford University Press, 1993).
34. V. I. Lenin, "Our Foreign and Domestic Position and Party Tasks," https://www.marxists.org/archive/lenin/works/1920/nov/21.htm.
35. C. Munn, *Anglo-China: Chinese People and British Rule in Hong Kong, 1841–1880* (Richmond, UK: Curzon, 2001), 67–98, 109–59, 257–89, 339–66. Pirates were also a threat on land as well as at sea as late as the twentieth century. In 1912 some forty to fifty pirates attacked Cheung Chau, killing three policemen. Strickland, *Southern District Officer Reports*, 177–81.
36. Munn, *Anglo-China*, 131–32.
37. W. Schivelbusch, *Disenchanted Night: The Industrialization of Light in the Nineteenth Century* (Berkeley: University of California Press, 1995), 82.
38. CO, *Hong Kong: Original Correspondence. 1841–1854*, vol. 1, 129. Austin Coates notes that lighting was made mandatory that same year, 1847, with all houses in non-Chinese areas required to display a lamp, typically fueled by peanut oil, over

1. PRIVATE LIGHT AND COLONIAL POWER

the main entrance. Coates, *A Mountain of Light: The Story of the Hongkong Electric Company* (London: Heinemann, 1977), 16.
39. Endacott, *A History of Hong Kong*, 70.
40. *The Hong Kong Government Gazette*, December 5, 1857, http://www.grs.gov.hk/ws/rhk/en/1850s.html.
41. Schivelbusch, *Disenchanted Night*, 95. Goh Chor Boon notes that in 1860s Singapore the brightest gas street lamps provided less light than a modern 25-watt bulb. Boon, *Technology and Entrepôt Colonialism in Singapore, 1819–1940* (Singapore: Institute of Southeast Asian Studies, 2013), 107.
42. Conversely, CLP, a corporate body, was sometimes envisioned in personal terms. District officer Austin Coates reported in the 1950s that he was "approaching Messrs [sic] China Light and Power regarding a supply of electric light." Strickland, *Southern District Officer Reports*, 217.
43. Hong Kong Government, *Report of the Working Party of Electrification of Squatter Areas* (Hong Kong: Government Printer, 1976).
44. Electricity meters allowed outside interests to gain privileged access to private dwellings. Tanja Winther describes the delicate relationship in Zanzibar between meter readers and customers as well as the perception that meter readers were the eyes of the state. Winther, *The Impact of Electricity: Development, Desires and Dilemmas* (New York: Berghahn Books, 2008), 108–11, 233.
45. S. T. Kwok, *A Century of Light* (Hong Kong: CLP, 2001), 92.
46. J. M. Carroll, *A Concise History of Hong Kong* (Hong Kong: Hong Kong University Press, 2007), 144.
47. HKG, *Report on the Riots in Kowloon and Tsuen Wan, October 10th to 12th, 1956, Together with Covering Despatch Dated the 23rd December, 1956, from the Governor of Hong Kong to the Secretary of State for the Colonies* (Hong Kong: Government Printer, 1956). The report gives the figure of fifty-nine deaths (44–46) and sixty deaths (paragraph seven of Grantham's covering letter). Grantham's letter, which was written after the main body of the report, suggests he had access to more up-to-date information. Adding to the uncertainty about the actual death toll, in his autobiography Grantham wrote that "more than 60" were killed. Grantham, *Via Ports*, 191–92.
48. I use the term "refugee," but the Hong Kong government did not. Former governor Robert Black wrote: "Hong Kong banished the word 'refugee.'" See R. B. Black, *Immigration and Social Planning in Hong Kong* (London: China Society, 1965), 7.
49. R. Hyam and W. R. Louis, *The Conservative Government and the End of Empire, 1957–1964, Part I: High Policy, Political and Constitutional Change* (London: The Stationery Office, 2000), 23.
50. CLP, Annual Report 1957, 4.
51. A. Allsop, ed., *St. George's Building: A Brief Portrait* (Hong Kong: Hong Kong Heritage Project, 2016), 97; HKHP, SEK-1A-055: 22. Kadoorie's nighttime illuminations were intended to promote a feeling of calm and security in a city where the absence of light could provoke panic. A switchboard failure at HKE's North Point plant in 1967 led to a blackout. In a jittery colony the incident had a "worse effect than all terrorist incidents put together." See "The Lights Went Out," *South China Morning Post*, October 28, 1967.
52. Kwok, *A Century of Light*, 90.

53. S. Ford, " 'Reel Sisters' and Other Diplomacy: Cathay Studios and Cold War Cultural Production," in *Hong Kong in the Cold War*, ed. P. Roberts and J. Carroll (Hong Kong: Hong Kong University Press, 2016), 194–95.
54. CLP, "Chairman's Statement to Shareholders, To be submitted at an Extraordinary General Meeting on 7th June 1962."
55. Hughes, *Networks of Power*, 15.

2. IN THE BEGINNING

1. D. E. Nye, *Electrifying America: Social Meanings of a New Technology, 1880–1940* (Cambridge, MA: MIT Press, 1990), 3–5; J. Jonnes, *Empires of Light: Edison, Tesla, Westinghouse, and the Race to Electrify the World* (New York: Random House, 2003), 330–33.
2. Hong Kong Government (hereafter HKG), Annual Report 2006, 421.
3. F. Bartlett, *The Peninsula: Portrait of a Grand Old Lady* (Hong Kong: Roundhouse Publications, 1997), 51; Lawrence Kadoorie, in *Tradition Well Served: The HSH Group & Kadoorie Family Documentary*, video (Hong Kong: Hongkong and Shanghai Hotels, n.d.).
4. Hong Kong Heritage Project (hereafter HKHP), CLP-2-036: Kadoorie, February 26, 1940, 3; see also "Modern Power Plant Opened by Governor," *Hongkong Telegraph*, February 27, 1940.
5. HKG (1940), *Administration Reports for the Year 1939*: 30, 46.
6. S. C. Fan, *The Population of Hong Kong* (Paris: Committee for International Coordination of National Research in Demography, 1974), 2.
7. A. Smart, *The Shek Kip Mei Myth: Squatters, Fires and Colonial Rule in Hong Kong, 1950–1963* (Hong Kong: University of Hong Kong Press, 2006), 123.
8. S. Selwyn-Clarke, *Footprints: The Memoirs of Sir Selwyn Selwyn-Clarke* (Hong Kong: Sino-American Publishing, 1975), 59–60.
9. P. Snow, *The Fall of Hong Kong* (New Haven, CT: Yale University Press, 2003), 42.
10. The Board decided on the 12.5 megawatt Hok Un expansion in 1935, three years after the Kadoories took control from Shewan Tomes. China Light & Power (hereafter CLP), Annual Report 1935.
11. A. Wright and H. A. Cartwright, eds., *Twentieth Century Impressions of Hongkong, Shanghai, and Other Treaty Ports of China: Their History, People, Commerce, Industries, and Resources* (London: Lloyd's Greater Britain Publishing, 1908); J. Wong and C. K. Wong, *China's Power Sector* (Singapore: World Scientific Publishing, 1999), 13–14.
12. J. Osterhammel and N. P. Petersson, *Globalization: A Short History* (Princeton, NJ: Princeton University Press, 2005), ix.
13. W. J. Hausman, P. Hertner, and M. Wilkins, *Global Electrification: Multinational Enterprise and International Finance in the History of Light and Power, 1878–2007* (Cambridge: Cambridge University Press, 2008); T. Hughes, *Networks of Power: Electrification in Western Society, 1880–1930* (Baltimore, MD: Johns Hopkins University Press, 1983), and *Human-Built World: How to Think About Technology and Culture* (Chicago: University of Chicago Press, 2004); J. Jonnes, *Empires of Light*; J. Lambert, *The Power Brokers: The Struggle to Shape and Control the Electric*

2. IN THE BEGINNING

Power Industry (Cambridge, MA: MIT Press, 2015); F. McDonald, *Insull* (Chicago: University of Chicago Press, 1962); Nye, *Electrifying America*; C. R. Roach, *Simply Electrifying: The Technology That Transformed the World, from Benjamin Franklin to Elon Musk* (Dallas, TX: Ben Bella Books, 2017).

14. D. Read, *The Power of News: The History of Reuters* (Oxford: Oxford University Press, 1999), 61.
15. Singapore introduced gas lighting May 24, 1864, to celebrate Queen Victoria's birthday but was relatively late to adopt distributed electricity. Street lighting started in 1906; as late as the 1920s, electricity was a rarity. G. C. Boon, *Technology and Entrepôt Colonialism in Singapore, 1819–1940* (Singapore: Institute of Southeast Asian Studies 2013), 107–9.
16. Hong Kong and China Gas Co., *Lighting the Past, Brightening the Future* (Hong Kong: Hong Kong and China Gas Co., 2012), 24.
17. CLP, Annual Report 1903.
18. P. Cunich, *A History of the University of Hong Kong* (Hong Kong: Hong Kong University Press, 2012), 217–19, 271–75, 327–28.
19. CLP, Annual Report 1957, 8.
20. HKHP, SEK-04-134, BR4/98.
21. G. W. Des Voeux, *My Colonial Service in British Guiana, St. Lucia, Trinidad, Fiji, Australia, Newfoundland, and Hong Kong, with Interludes* (London: John Murray, 1903), 2:206.
22. T. S. Pugh, "Telephones," in *Hong Kong Business Symposium: A Compilation of Authoritative Views on the Administration, Commerce and Resources of Britain's Far East Outpost*, ed. J. M. Braga (Hong Kong: South China Morning Post, 1957), 316.
23. Read, *The Power of News*, 61; D. Winseck and R. Pike, *Communication and Empire: Media, Markets, and Globalization, 1860–1930* (Durham, NC: Duke University Press, 2007), 3. Reuters made a payment of £3.9s to "Lambert of Hong Kong," "possibly an agent," in October 1862, suggesting that the agency may have had a presence in the colony as early as that date, notes Reuters archivist David Cutler, who provided a typewritten list of key dates in Hong Kong; this list cites the "cash book" as the source for the Lambert payment. Reuters started a Special India and China Service in 1859 (David Cutler, personal communication).
24. For the claim that the Peak Tram was Asia's first cable railway (properly speaking, it is a funicular), see Hongkong and Shanghai Hotels, *Tradition Well Served: The HSH Group & Kadoorie Family Documentary* (video, 2014).
25. I. B. Trevor, "The Kowloon-Canton Railway," in Braga, *Hong Kong Business Symposium*, 325.
26. HKG, Civil Aviation Department, Kai Tak Airport (1925–1998), https://www.cad.gov.hk/english/kaitak.html.
27. J. G. O'Donnell, "Pan American's 'Story in the Pacific,'" in Braga, *Hong Kong Business Symposium*, 128; G. B. Endacott, *A History of Hong Kong* (Hong Kong: Oxford University Press,1973), 294.
28. HKG, Annual Report 1928, 13.
29. Des Voeux, *My Colonial Service*, 2:228.
30. Des Voeux, *My Colonial Service*, 2:203.
31. Des Voeux, *My Colonial Service*, 2:249–52.
32. Endacott, *A History of Hong Kong*, 272.

2. IN THE BEGINNING

33. A. Walker and S. Rowlinson, *The Building of Hong Kong: Constructing Hong Kong Through the Ages* (Hong Kong: Hong Kong University Press, 1990), 113–14.
34. I. Lambot and G. Chambers, *One Queen's Road Central* (Hong Kong: Hongkong Bank, 1986), 58–87; Walker and Rowlinson, *The Building of Hong Kong*, 60–61.
35. Lambot and Chambers, *One Queen's Road Central*, 80.
36. J. F. Tsai, *Hong Kong in Chinese History: Community and Social Unrest in the British Colony, 1842–1913* (New York: Columbia University Press, 1993), 65–102, 124–46, 147–81, 182–206, 288–96. This issue is also explored in chapter 3 in a discussion of failed political reforms proposed by Governor Mark Young.
37. L. Hannah, *British Electricity Before Nationalisation: A Study of the Development of the Electricity Supply Industry in Britain to 1948* (London: Macmillan, 1979), 4. Britain's first central electricity supply station was established in 1881 at Godalming, in Surrey, by Siemens. The water-powered facility was used for street lighting and private houses. Inadequate demand led to the restoration of gas lighting in 1884 (7).
38. Des Voeux, *My Colonial Service*, 2:206.
39. Des Voeux, *My Colonial Service*, 2:312.
40. Powercor Australia: "Electricity in Early Victoria and Through the Years," https://www.powercor.com.au/media/1251/fact-sheet-electricity-in-early-victoria-and-through-the-years.pdf; The Brihanmumbai Electric Supply & Transport Undertaking, "Electricity Arrives in Mumbai," https://web.archive.org/web/20080917080729/http://www.bestundertaking.com:80/his_chap04.asp.
41. C. Hou, *Foreign Investment and Economic Development in China: 1840–1937* (New York: Routledge, 2000), 89, 247n107.
42. Wright and Cartwright, *Twentieth Century Impressions*, 371, 395–98, 405, 432, 438.
43. Wright and Cartwright, *Twentieth Century Impressions*, 594, 668, 694, 708, 730, 738, 756, 788, 812, 830, 836.
44. D. G. Victor and T. C. Heller, eds., *The Political Economy of Power Sector Reform: The Experiences of Five Major Developing Countries* (Cambridge: Cambridge University Press, 2007), 81.
45. Wright and Cartwright, *Twentieth Century Impressions*, 594.
46. CESC Limited, "The Story of Electricity in the City of Kolkata," https://htshift.jimdo.com/cesc-limited-a-nice-journey/.
47. E. E. Patalinghug, "An Analysis of the Philippine Electric Power Industry,'" 2003, http://www.ombudsman.gov.ph/UNDP4/wp-content/uploads/2013/01/An-Analysis-of-the-Philippine-Electric_Patilinhug.pdf; Meralco company website, http://www.meralco.com.ph/about-us/history. In a reminder of the capital-intensive nature of electricity supply companies, and in a foreshadowing of remarks made about Esso's subsequent investment in CLP, Meralco's corporate history notes: "The facilities that Meralco built to provide these two services represented for many years the largest single investment of American private capital and know-how in the whole of East Asia."
48. Ryrie was the first chairman of the Hong Kong Jockey Club, a founding investor in the Peak Tram, a three-time chairman of the Hong Kong General Chamber of Commerce, and an investor in Dairy Farm. He was an Unofficial Member of the Legislative Council. He had earlier been senior partner at opium merchant Turner & Co. M. Holdsworth and C. Munn, eds., *Dictionary of Hong Kong Biography* (Hong Kong: Hong Kong University Press, 2012), 379–80.

2. IN THE BEGINNING

49. R. Hutcheon, *The Blue Flame: 125 Years of Towngas in Hong Kong* (Hong Kong: Hong Kong & China Gas, 1987), 25.
50. *Hongkong Telegraph*, January 25, 1889.
51. A. Coates, *A Mountain of Light: The Story of the Hongkong Electric Company* (London: Heinemann, 1977), 6–7; *Hongkong Telegraph*, January 25, 1889.
52. *Hongkong Telegraph*, January 25, 1889. The article is unsigned, but articles from the newspaper during this period are generally attributed to Fraser-Smith, who founded the newspaper in 1881 and controlled it until his death in 1895.
53. Coates, *A Mountain of Light*, 40.
54. Coates, *A Mountain of Light*, 35–36. Endacott, *A History of Hong Kong* (259), cites an 1890 "reproof" by the secretary of state in connection with street lighting.
55. *Hongkong Weekly Press and China Overland Trade Report* (hereafter *HWP*), November 26, 1898, 438.
56. The documents establishing Hong Kong as a Crown Colony and setting up its administration were contained in the Chartered Letters Patent and other royal instructions dated April 5, 1843. C. Collins, *Public Administration in Hong Kong* (London: Royal Institute of International Affairs, 1952), 46.
57. Wright and Cartwright, *Twentieth Century Impressions*, 209.
58. HKHP, CLP-03A-09, Noel Croucher interview, March 27, 1979.
59. Wright and Cartwright, *Twentieth Century Impressions*, 213, 790, 792.
60. *HWP*, November 26, 1898, 438; see also Wright and Cartwright, *Twentieth Century Impressions*, 792.
61. N. Cameron, *Power: The Story of China Light* (Hong Kong: Oxford University Press, 1982), 18–19, 27.
62. L. Kadoorie, *Chairman's Speech for Inaugural Ceremony of New Power Station* (Hong Kong: CLP, 1940). See also HKHP, CLP-03A-09, Letter from I. W. Shewan, Robert Shewan's nephew, to Horace Kadoorie, September 13, 1949.
63. L. Kadoorie, "China Light & Power Co. Ltd.: 1901–1918" (Hong Kong: CLP, 1977).
64. *HWP*, June 2, 1902, 418.
65. HKHP, CLP-3A-09, Letter from I. W. Shewan.
66. HKHP, *One Hundred Years of Kadoorie Business Records at a Glance* (Hong Kong: Hong Kong Heritage Project, 2011), 6. See also Y. C. Kong, "Jewish Merchants' Community in Shanghai: A Study of the Kadoorie Enterprise, 1890–1950" (PhD dissertation, Hong Kong Baptist University, 2017), 176–207.
67. Chi Zhang and Thomas C. Heller, "Reform of the Chinese Electric Power Market: Economics and Institutions" in Victor and Heller, *The Political Economy of Power Sector Reform*, 81.
68. *HWP*, January 27, 1890; *Hongkong Telegraph*, February 10, 1890, and April 8, 1890.
69. Jonnes, *Empires of Light*, 220.
70. *HWP*, January 27, 1890: 2; *Hongkong Telegraph*, February 10, 1890, and April 8, 1890.
71. Kadoorie, "China Light & Power Co.," 2.
72. Wright and Cartwright, *Twentieth Century Impressions*, 792.
73. F. Dikötter, *Exotic Commodities: Modern Objects and Everyday Life in China* (New York: Columbia University Press, 2006), 141, 182–84.
74. R. A. Caro, *The Path to Power* (New York: Knopf, 1982), 525.
75. Wright and Cartwright, *Twentieth Century Impressions*, 790.
76. Wright and Cartwright, *Twentieth Century Impressions*, 792.

2. IN THE BEGINNING

77. Kadoorie, "China Light & Power Co.," 2.
78. CLP, Annual Report 1903.
79. Kadoorie, "China Light & Power Co.," 1.
80. E. Sinn, *Pacific Crossing: California Gold, Chinese Migration, and the Making of Hong Kong* (Hong Kong: Hong Kong University Press, 2014), 134.
81. Kadoorie, "China Light & Power Co.," 2.
82. Cameron, *Power*, 17; J. Bradley, *The Imperial Cruise: A Secret History of Empire and War* (New York: Little, Brown, 2009), 290–97.
83. Tsai, *Hong Kong in Chinese History*, 188, 190.
84. Kadoorie, "China Light & Power Co.," 6.
85. HKHP, CLP-3A-01, and CLP-03-099.
86. Endacott, *A History of Hong Kong*, 276. By 1916 CLP management was satisfied enough to write that "we see no reason to change our opinion as to the remunerative nature of the contract we have made to supply our neighbors, the Hongkong and Whampoa Dock Co." (CLP, Annual Report 1916).
87. *HWP*, February 14, 1903, and February 28, 1903. Factories needed to be reengineered to accommodate electricity. This slowed the adoption of electricity, notably in U.S. factories.
88. CLP, *Submissions and Final Address of the China Light & Power Co., Ltd., to the Electricity Supply Companies Commission 1959* (Hong Kong: CLP, 1959), 6.
89. McDonald, *Insull*, 55–132.
90. *HWP*, June 2, 1902.
91. *HWP*, April 25, 1903.
92. *HWP*, February 14, 1903, and subsequent issues in February, March, and April.
93. HKHP, CLP 3A-2: "Sir Lawrence Kadoorie's Address to the Staff on 8th March 1977 at the Peninsula Hotel." The Westinghouse equipment appears to have been British. See C. Stratford, "New Kowloon Power Station of the China Light and Power Co., (1918) Ltd.," *The Far Eastern Review*, June 1931, 367.
94. CLP, Annual Report 1913.
95. HKHP, CLP-3A-09: 1 (Noel Braga, 1976). Noel Braga first appears as company secretary in CLP printed records in an extraordinary general meeting held on April 23, 1928. This is the same year that his father, J. P. Braga, and Elly Kadoorie joined the Board and also the year that they convinced Governor Cecil Clementi to give CLP the right to electrify the New Territories. Both Braga and Kadoorie were appointed to the board during the course of the year but were only elected at the December 24, 1928, Ordinary Yearly Meeting (CLP, Annual Report 1928).
96. HKHP, CLP-3A-09, 1 (Noel Braga, 1976).
97. Wright and Cartwright, *Twentieth Century Impressions*, 131.
98. Wright and Cartwright, *Twentieth Century Impressions*, 166.
99. D. Waters, "Hong Kong Hongs with Long Histories and British Connections," *Journal of the Hong Kong Branch of the Royal Asiatic Society* 30 (1990): 248.
100. HKG, *Administrative Reports* (1928), Appendix Q: 25–27. These figures are for Hong Kong. In Kowloon, gas lamps declined by 2 to 534, "due to electric lights being substituted in Nathan Road."
101. S. T. Kwok, *A Century of Light* (Hong Kong: CLP, 2001), 9.
102. See, for example, HKG, *Administrative Reports* (1928): Annex A: Q 98–99.
103. Wright and Cartwright, *Twentieth Century Impressions*, 331.

2. IN THE BEGINNING

104. Wright and Cartwright, *Twentieth Century Impressions*, 162–63, 169.
105. Wright and Cartwright, *Twentieth Century Impressions*, 331; S. Vines, *The Story of St John's Cathedral* (Hong Kong: FormAsia, 2001), 48.
106. B. E. Baker, *Shanghai: Electric and Lurid City: An Anthology* (Hong Kong: Oxford University Press, 1998), 72.
107. Wright and Cartwright, *Twentieth Century Impressions*, 830, 836.
108. HKHP, CLP-3A-09: 2 (Noel Braga, 1976).
109. CLP, Annual Report 1917, 1920.
110. CLP, *A Century of Power* (director's cut) (video, 2017).
111. Wong Siu-lun discusses the creation of "system trust" in Hong Kong "through the stable sociopolitical and economic framework afforded by British colonialism." See S. Wong, "Chinese Entrepreneurs and Business Trust," in *Business Networks and Economic Development in East and Southeast Asia*, ed. G. G. Hamilton (Hong Kong: University of Hong Kong, 1996), 23. At the time of Kadoorie's naturalization in 1927, that sense of a colonially created system trust would have extended to Shanghai as well. Although Wong focuses on Chinese entrepreneurs, the Baghdadi Jewish business community in Hong Kong and Shanghai exemplified the benefits of system trust. See M. J. Meyer, *From the Rivers of Babylon to the Whangpoo: A Century of Sephardi Jewish Life in Shanghai* (Lanham, MD: University Press of America, 2003), and *Shanghai's Baghdadi Jews: A Collection of Biographical Reflections* (Hong Kong: Blacksmith Books, 2015). See also Kong, "Jewish Merchants' Community."
112. Kadoorie's quest for British nationality must also be seen in the context of Baghdadi Jews in East Asia after 1906 "being denied or stripped of their protection" and in Iraq, after World War I, "where many Jews traumatized by the First World War and fearful of Iraq's postwar fate sought but were denied British naturalization." Sarah Abrevaya Stein, "Protected Persons? The Baghdadi Jewish Diaspora, the British State, and the Persistence of Empire," *American Historical Review* 116, no. 1 (February 2011): 86.
113. CLP, Annual Report 1928.
114. Kong, "Jewish Merchants' Community," details the Kadoorie family's interests in Shanghai during this period.
115. Cameron, *Power*, 95.
116. Lawrence Kadoorie, "The Kadoorie Memoir," in D. Leventhal and Lawrence Kadoorie, *Sino-Judaic Studies: Whence and Whither: An Essay and Bibliography and the Kadoorie Memoir* (Hong Kong: Hong Kong Jewish Chronicle, 1985), 91.
117. HKHP, CLP-03-075 (1928–33, especially the extraordinary general meeting [EGM] of March 2, 1932), CLP-03-088, and CLP-02-036. Unusually, the EGM was not held at CLP's head office, which was Shewan Tomes's, but at the nearby offices of accountants Lowe Bingham & Matthews. Lawrence Kadoorie and J. P. Braga attended the EGM; Elly Kadoorie did not. Elly's first recorded appearance at an official CLP event was the ordinary yearly meeting in December 1932.
118. The CLP board takeover was unusual but not unique for the Kadoories. Horace Kadoorie was elected as chairman of Shanghai Gas in a contested election in 1939 following a campaign against incumbent chairman L. E. Canning. Kong, "Jewish Merchants' Community," 195–99.
119. F. H. H. King, *The Hongkong Bank in Late Imperial China, 1864–1902* (Cambridge: Cambridge University Press, 1987), 462–66.

3. WAR, OCCUPATION, AND NEW POSSIBILITIES

120. Allsop, *St. George's Building*, 2016, 31; "Death of Mr. R. G. Shewan," *Hong Kong Daily Press*, February 16, 1934. The *Daily Press* reported that Shewan "met his death on Wednesday morning under tragic circumstances. His body was found at 6.55 a.m. in the grounds of his house at 2 Conduit Road and it is believed that he fell from his window while leaning out."
121. HKHP, Diaries (Book 9): SEK-1A-001 A01/08.
122. The document continues to resonate through the company. Unprompted, in 2016 CLP's CEO Richard Lancaster mentioned it and volunteered information about its importance (personal communication).
123. L. Kadoorie, "General Policy (A Brief Survey)," 2.
124. Kadoorie, "General Policy," 3.
125. Cameron, *Power*, 242; Coates, *A Mountain of Light*, 194.
126. Kadoorie, "General Policy," 6.
127. Kadoorie, "General Policy," 1.
128. Kadoorie, "General Policy," 3.
129. Kadoorie, "General Policy," 6.
130. Kadoorie, "General Policy," 9.
131. Kadoorie, "General Policy," 6.
132. Kadoorie, "General Policy," 6.
133. Kadoorie, "General Policy," 9.
134. HKHP, Diaries (Book 9): SEK-1A-001 A01/08.
135. Wright and Cartwright, *Twentieth Century Impressions*, 156.
136. "Modern Power Plant Opened by Governor," *Hongkong Telegraph*, February 27, 1940.
137. "Modern Power Plant Opened by Governor," *Hongkong Telegraph*.
138. HKHP, SEK-04-132.
139. HKHP, SEK-04-136.
140. HKHP, SEK-04-139.
141. "Modern Power Plant Opened by Governor," *Hongkong Telegraph*.
142. "Modern Power Plant Opened by Governor," *Hongkong Telegraph*.
143. "Modern Power Plant Opened by Governor," *Hongkong Telegraph*.
144. CLP, Annual Report 1937.
145. HKHP, SEK-04-135: DRI/81.
146. For a fuller analysis of these issues, see Stein, "Protected Persons?," 80–108.

3. WAR, OCCUPATION, AND NEW POSSIBILITIES

1. P. Snow, *The Fall of Hong Kong* (New Haven, CT: Yale University Press, 2003), 51.
2. Quoted in G. B. Endacott, *Hong Kong Eclipse* (Hong Kong: Oxford University Press, 1978), 56.
3. Snow, *The Fall of Hong Kong*, 50–52.
4. S. Selwyn-Clarke, *Footprints: The Memoirs of Sir Selwyn Selwyn-Clarke* (Hong Kong: Sino-American Publishing, 1975), 58.
5. Snow, *The Fall of Hong Kong*, 348.
6. A. Walker and S. Rowlinson, *The Building of Hong Kong: Constructing Hong Kong Through the Ages* (Hong Kong: Hong Kong University Press, 1990), 204.

3. WAR, OCCUPATION, AND NEW POSSIBILITIES

7. S. Tsang, *Governing Hong Kong: Administrative Officers from the Nineteenth Century to the Handover to China, 1862–1997* (Hong Kong: Hong Kong University Press, 2007), 146; N. Monnery, *Architect of Prosperity: Sir John Cowperthwaite and the Making of Hong Kong* (London: London Publishing Partnership, 2017), 74–78. For details of the 1940 property, salaries, and profits taxes, see Endacott, *Hong Kong Eclipse*, 40–41.
8. E. F. Szczepanik, *The Economic Growth of Hong Kong* (London: Oxford University Press, 1958), 66–67.
9. Shu Ching Fan, *The Population of Hong Kong* (Paris: Committee for International Coordination of National Research in Demography, 1974), 2. Note that Fan uses an end-war population figure of 600,000, the upper end of the general range of 500,000 to 600,000; I use 500,000 in calculating population growth.
10. Fan, *The Population of Hong Kong*, 8.
11. CLP, Annual Report 1955, 11.
12. HKHP, SEK-1A-07 (diary, July 17, 1941). See also CLP, *Submissions and Final Address of the China Light & Power Co., Ltd., to the Electricity Supply Companies Commission 1959* (Hong Kong: CLP, 1959), 40 (hereafter *Submissions*). The use of the surcharge to boost profits was apparent from the beginning. In a diary entry of June 27, 1941, Horace wrote Lawrence: "Father wishes to know whether the surcharge of 10% is now in force and from what date this took place. Is this surcharge sufficient to assure our usual dividend? Please let us know" (HKHP, SEK-1A-07).
13. A. Gerschenkron, *Economic Backwardness in Historical Perspective* (Cambridge, MA: Harvard University Press, 1962).
14. C. Mosk, *Japanese Industrial History: Technology, Urbanization, and Economic Growth* (Armonk, NY: M. E. Sharpe, 2001).
15. A. S. Moore, *Constructing East Asia: Technology, Ideology, and Empire in Japan's Wartime Era, 1931–1945* (Stanford, CA: Stanford University Press, 2013), 48–52. Aikawa recognized that the construction of these hydropower projects would require the large-scale use of labor; he saw human labor as just another sort of energy, what he called "blood gasoline" (52).
16. D. Fortescue, *The Survivors: A Period Piece* (London: Anima Books, 2015), 115.
17. HKHP, SEK-1A-055: 2.
18. S. T. Kwok, *A Century of Light* (Hong Kong: CLP, 2001), 48.
19. Endacott, *Hong Kong Eclipse*, 77.
20. Snow, *The Fall of Hong Kong*, 56–57.
21. N. Cameron, *Power: The Story of China Light* (Hong Kong: Oxford University Press, 1982), 141.
22. HKHP, CLP-03A-09, October 19, 1979, interview.
23. CLP, *Submissions*, 2.
24. G. B. Endacott, *A History of Hong Kong* (Hong Kong: Oxford University Press, 1973), 300.
25. C. M. Kwong and Y. L. Tsoi, *Eastern Fortress: A Military History of Hong Kong, 1840–1970* (Hong Kong: University of Hong Kong Press, 2014), 196–97.
26. The following section draws on A. Coates, *A Mountain of Light: The Story of the Hongkong Electric Company* (London: Heinemann, 1977), 144–52, and Endacott, *Hong Kong Eclipse*, 85, 88.
27. Coates, *A Mountain of Light*, 147.

3. WAR, OCCUPATION, AND NEW POSSIBILITIES

28. Government of the United Kingdom, *The War Dead of the British Commonwealth and Empire: The Register of the Names of Those Who Fell in the 1939–1945 War and Are Buried in Cemeteries in Hong Kong* (London: Imperial War Graves Commission, 1956), 10.
29. HKHP, CLP-03A-09, "Extracts Board Mtg. Minutes Sep. 45–Apr. 53."
30. CLP, *Submissions*, 2.
31. HKHP, CLP-03A-09, "Extracts Board Mtg. Minutes Sep. 45–Apr. 53"; Cameron, *Power*, 150–152.
32. Endacott, *Hong Kong Eclipse*, 115.
33. Coates, *A Mountain of Light*, 104, 152.
34. Coates, *A Mountain of Light*, 152.
35. P. Cunich, *A History of the University of Hong Kong* (Hong Kong: Hong Kong University Press, 2012), 427–28.
36. Endacott, *Hong Kong Eclipse*, 128.
37. Endacott, *Hong Kong Eclipse*, 147.
38. Kwok, *A Century of Light*, 48.
39. Cameron, *Power*, 147; Kwok, 2001: 48.
40. S. F. Li, *Hong Kong Surgeon* (London: Victor Gollancz, 1964), 137.
41. Li, *Hong Kong Surgeon*, 137 (emphasis in the original).
42. Li, *Hong Kong Surgeon*, 137.
43. Kwok, *A Century of Light*, 50.
44. Endacott, *Hong Kong Eclipse*, 149.
45. Kwok, *A Century of Light*, 52.
46. Cameron, *Power*, 268.
47. CLP, Annual Reports 1903, 1906, 1907, 1916.
48. G. C. Emerson, *Hong Kong Internment, 1942 to 1945: Life in the Japanese Civilian Camp at Stanley* (Hong Kong: Hong Kong University Press, 2008), 86–87.
49. "21 August 1944 Chronology of Events Related to Stanley Civilian Internment Camp," https://gwulo.com. Cunich reports that nutritional deficiency diseases such as beriberi and pellagra begin to affect residents of Stanley Camp from March 1942 on. Cunich, *History of the University of Hong Kong*, 406–7.
50. B. Anslow, *Tin Hats & Rice: A Diary of Life as a Hong Kong Prisoner of War, 1941–1945* (Hong Kong: Blacksmith Books, 2018), 181.
51. Raymond E. Jones, diary, August 22, 1944 (hereafter Jones diary). Transcription of diary at https://gwulo.com/node/19934.
52. Jones diary, August 22, 1944.
53. Kwok, *A Century of Light*, 54.
54. Harry Ching, diary, July 1944. Transcription of diary at https://gwulo.com/node/14227.
55. P. Liberman, *Does Conquest Pay? The Exploitation of Occupied Industrial Societies* (Princeton, NJ: Princeton University Press, 1996), 103. The shortfall in Hong Kong occurred despite increased coal production in Korea, Taiwan, and Manchukuo; in 1941–1944, annual coal output from these areas was nearly double the 1937 level (106).
56. Anslow, *Tin Hats & Rice*, 253–55.
57. HKHP, CLP-3A-09: Stephens letter.
58. W. R. Louis, "Hong Kong: The Critical Phase, 1945–1949," *American Historical Review* 102, no. 4 (October 1997): 1062.

3. WAR, OCCUPATION, AND NEW POSSIBILITIES

59. A. J. Whitfield, *Hong Kong, Empire and the Anglo-American Alliance at War, 1941–45* (Hong Kong: Hong Kong University Press, 2001), 113–15.
60. C. Loh, *Underground Front: The Chinese Communist Party in Hong Kong* (Hong Kong: Hong Kong University Press, 2010), 63–66; S. Tsang, *A Modern History of Hong Kong* (Hong Kong: Hong Kong University Press, 2004), 134–38; J. M. Carroll, *A Concise History of Hong Kong* (Hong Kong: Hong Kong University Press, 2007), 127. During the war, Gimson opposed calls by Stanley prisoners for repatriation on the grounds that sending British detainees out of Stanley would make it harder for Britain to regain sovereignty at the end of the war. F. Gimson, "Internment in Hongkong" (unpublished manuscript), appendix 1; Whitfield, *Hong Kong*, 203).
61. Endacott, *Hong Kong Eclipse*, 257, 262–64.
62. Tsang, *Governing Hong Kong*, 52; E. Oxford, *At Least We Lived: The Unlikely Adventures of an English Couple in World War II China* (Charleston, SC: Branksome Books, 2013), 82–98.
63. Szczepanik, *Economic Growth*, 65–66.
64. F. J. Wakefield, "Hong Kong Coaling Agency," in *Hong Kong Business Symposium: A Compilation of Authoritative Views on the Administration, Commerce and Resources of Britain's Far East Outpost*, ed. J. M. Braga (Hong Kong: South China Morning Post, 1957), 407.
65. Szczepanik, *Economic Growth*, 16–17.
66. Szczepanik, *Economic Growth*, 34–39.
67. Tsang, *Modern History of Hong Kong*, 139, 140.
68. Tsang, *Modern History of Hong Kong*, 140.
69. Tsang, *Modern History of Hong Kong*, 138.
70. Fortescue, *The Survivors*, 122.
71. Snow, *The Fall of Hong Kong*, 259.
72. Tsang, *Modern History of Hong Kong*, 141–44 and Carroll, *A Concise History of Hong Kong*, 130–32.
73. Snow, *The Fall of Hong Kong*, 184.
74. A. W. Grantham, *Via Ports, from Hong Kong to Hong Kong* (Hong Kong: Hong Kong University Press, 1965), 104.
75. V. England, *Kindred Spirits: A History of the Hong Kong Club* (Hong Kong: Hong Kong Club, 2016), 110–14.
76. England, *Kindred Spirits*, 64. See also China Mail, *Who's Who in the Far East, 1906-7* (Hong Kong: China Mail, 1906), 174.
77. England, *Kindred Spirits*, 64–65.
78. Hong Kong Club, personal communication.
79. Kadoorie, "The Kadoorie Memoir," in D. Leventhal and Lawrence Kadoorie, *Sino-Judaic Studies: Whence and Whither: An Essay and Bibliography and the Kadoorie Memoir* (Hong Kong: Hong Kong Jewish Chronicle, 1985), 95.
80. CLP, *Submission*, 2.
81. CLP, *A Century of Power*, video.
82. HKHP, SEK-3A-064, R.2-A-1/2.
83. HKHP, CLP-3A-09, Letter to Stephens.
84. HKHP, CLP-3A-09, Letter to Stephens.
85. Videotaped interviews with Michael Kadoorie, June 3 and December 16, 2008.
86. CLP, *A Century of Power*, video.

87. HKHP, SEK-3A-065, R2-B1/15.
88. "Monopoly Profits Denied," *China Mail*, October 22, 1959.
89. Cameron, *Power*, 168.
90. HKHP, SEK-1A-055, 14.
91. The price of coal was "a great deal higher" than before the war, stores and materials costs were four to six times higher than prewar levels, and wages had trebled. Hong Kong Government, Annual Report 1946, 81.
92. HKHP, SEK-3A-065, R.2-B-1.
93. HKHP, CLP-2-119.
94. HKHP, SEK-3A-065, R.2-B-1/40.
95. HKHP, SEK-3A-065, R.2-B-1/ 32–33, 42, 46, 48, 50, 50a, 51, 53–56.
96. Paul Steinschneider was Austrian, and he arrived in Hong Kong in mid-1939 after initially fleeing Austria to Shanghai. He was hired as an engineer by the Hong Kong and Whampoa Dock Co. partly on humanitarian grounds (Jewish refugees had to acquire visas, employment, and a guarantee that they would not be a charge on public funds to ensure they could stay in the colony). Personal communication, Amelia Allsop, 2017.
97. Cameron, *Power*, 154.
98. "Edgar Laufer," A Borrowed Place: Jewish Refugees in Hong Kong, 1938–1953, https://hongkongrefuge.wordpress.com/2016/08/21/edgar-laufer/.
99. Cameron, *Power*, 155.
100. HKHP, CLP-2-036.
101. E. G. Pryor, *Housing in Hong Kong* (Hong Kong: Oxford University Press, 1983), 26, 30; Urban Renewal Authority (Hong Kong), 2018 Exhibition.
102. HKHP, CLP-03A-09, "Extract Board Mtg. Minutes, Sept. 45—Apr. 53"; Cameron, *Power*, 157. The September 1945 board meeting held a silent tribute to those who had died during the war, including Elly Kadoorie, J. P. Braga, and the CLP personnel who were killed in the fighting for Hong Kong.
103. CLP, *Submission*, 2.
104. Videotaped interviews with Michael Kadoorie, 2016.
105. F. H. H. King, *The Hongkong Bank in the Period of Development and Nationalism, 1941–1984* (Cambridge: Cambridge University Press, 1991), 55.
106. King, *Hongkong Bank*, 87, 938n22.
107. HKHP, CLP-3A-09. King notes that CLP unsuccessfully tried to cancel all or part of this order on Sept. 28, 1945, just a few days after Lawrence Kadoorie wrote Stephens. King, *Hongkong Bank*, 87–88 and 938n22).
108. HKHP, CLP-3A-09.
109. HKHP, CLP-3A, C.2-D-1; this refers to a November 22, 1945, diary entry.
110. In mid-1960 Kadoorie suggested setting up a subsidiary company that could purchase power in bulk for sale to the People's Republic of China (HKHP, SEK-04-020, C.2-K-5/6, 17a).

4. "A PROBLEM OF PEOPLE"

1. J. Bew, *Clement Atlee: The Man Who Made Modern Britain* (Oxford: Oxford University Press, 2017).

4. "A PROBLEM OF PEOPLE"

2. L. Hannah, *British Electricity Before Nationalisation: A Study of the Development of the Electricity Supply Industry in Britain to 1948* (London: Macmillan, 1979).
3. Grantham wrote on June 21, 1949, that the Chinese Reform Association was formed as a "counterblast" to the elite expatriate-led Reform Association, and had "naturally attracted several political adventurers, notably Moscow-trained Percy Chen." Chen served on the Tenancy Tribunal Panel in 1948, when the colony's housing shortage led to the imposition of rent controls (Carl T. Smith collection, No. 2-403). During the 1967 disturbances, Chen (then the Hong Kong Chinese Reform Association chairman) was forced to go into hiding. Association Secretary Choi Wai-hung, a director of Hutchison International and one of the Association's key members, was detained in July 1967 for eighteen months without trial. The group was one of the few to hoist the PRC flag and distribute relief food and medicines donated by the PRC during the 1950s. G. K. Cheung, *Hong Kong's Watershed: The 1967 Riots* (Hong Kong: Hong Kong University Press, 2009), 204–13.
4. Hannah, *British Electricity*; T. Hughes, *Networks of Power: Electrification in Western Society, 1880–1930* (Baltimore, MD: Johns Hopkins University Press, 1983); W. J. Hausman, P. Hertner, and M. Wilkins, *Global Electrification: Multinational Enterprise and International Finance in the History of Light and Power, 1878–2007* (Cambridge: Cambridge University Press, 2008); S. S. Kale, *Electrifying India: Regional Political Economies of Development* (Stanford, CA: Stanford University Press, 2014); R. C. Tobey, *Technology as Freedom: The New Deal and the Electrical Modernization of the American Home* (Berkeley: University of California Press, 1996).
5. Tobey, *Technology as Freedom*, 194–214.
6. Hannah, *British Electricity*, 329–56.
7. Shu Ching Fan, *The Population of Hong Kong* (Paris: Committee for International Coordination of National Research in Demography, 1974), 2. See chapter 2, note 9, for my count of the population.
8. A. W. Grantham, *Via Ports, from Hong Kong to Hong Kong* (Hong Kong: Hong Kong University Press, 1965), 172.
9. R. Hyam and W. R. Louis, *The Conservative Government and the End of Empire, 1957–1964, Part II: Economics, International Relations, and the Commonwealth* (London: The Stationery Office, 2000), 729.
10. China Light & Power (hereafter CLP), *Submissions and Final Address of the China Light & Power Co., Ltd., to the Electricity Supply Companies Commission 1959* (Hong Kong: CLP, 1959), 5.
11. E. F. Szczepanik, *The Economic Growth of Hong Kong* (London: Oxford University Press, 1958), 45–57.
12. CLP, *Submissions*, 5.
13. A. W. Grantham, *Via Ports: From Hong Kong to Hong Kong*, rev. ed. (Hong Kong: Hong Kong University Press, 2012), 105.
14. Hong Kong Government (hereafter HKG), Annual Report 1956, 11.
15. J. M. Carroll, *A Concise History of Hong Kong* (Hong Kong: Hong Kong University Press, 2007), 142.
16. Grantham, *Via Ports*, 172.
17. C. R. Schenk, *Hong Kong as an International Financial Centre: Emergence and Development 1945–1965* (London: Routledge, 2001), 40–41.

4. "A PROBLEM OF PEOPLE"

18. M. Holdsworth and C. Munn, eds., *Dictionary of Hong Kong Biography* (Hong Kong: Hong Kong University Press, 2012), 53.
19. HKG, Annual Report 1956.
20. Sir Alexander Grantham, "Hong Kong," *Journal of the Royal Central Asian Society* 46, no. 2 (1959): 121.
21. Grantham, "Hong Kong," 123.
22. HKG, Annual Report 1959, 133–34.
23. HKG Annual Report 1956, 13.
24. HKG Annual Report 1956, 17.
25. The Shek Kip Mei fire is conventionally used as a starting point for the public housing program; in fact, Hong Kong's first public housing project, the Sheung Li Uk project in Sham Shui Po, was completed in 1952. Although the Shek Kip Mei fire spurred government action, the government's interest in providing housing predates the Japanese occupation. M. Jones, "Tuberculosis, Housing and the Colonial State: Hong Kong, 1900–1950," *Modern Asian Studies* 37, no. 3 (2003): 653–82. See also A. Smart, *The Shek Kip Mei Myth: Squatters, Fires and Colonial Rule in Hong Kong, 1950–1963* (Hong Kong: Hong Kong University Press, 2006).
26. HKG Annual Report 1956, 14.
27. HKG, Annual Report 1956, 27, 29.
28. HKG, *Report on the Riots in Kowloon and Tsuen Wan, October 10th to 12th, 1956, Together with Covering Despatch Dated the 23rd December, 1956, from the Governor of Hong Kong to the Secretary of State for the Colonies* (Hong Kong: Government Printer, 1956), 1–5; Grantham, *Via Ports*, 191–92. The HKG report gives a death toll of both fifty-nine (44–46) and sixty (paragraph seven of Grantham's covering letter).
29. HKG, *Report on the Riots*, 4.
30. Grantham, *Via Ports*, 191–92; HKG, *Report on the Riots*, 1–5.
31. H. J. Wiens, "Riverine and Coastal Junks in China's Commerce," *Economic Geography* 31, no. 3 (1955): 248–64.
32. E. Sinn, *Pacific Crossing: California Gold, Chinese Migration, and the Making of Hong Kong* (Hong Kong: Hong Kong University Press, 2014).
33. HKG, Hong Kong Legislative Council Proceedings, June 27, 1951: 185–86. Public Order Ordinance, 1948. Frontier Closed Order, 1951. Government. No. A. 100.
34. W. James, "Hong Kong—Boom Town," *Collier's*, January 10, 1948, 52–53.
35. As the United Press's Nanking bureau chief, James was aboard the *Panay* when the U.S. naval ship was sunk by Japanese bombers in December 1937; he was honored with the Navy Expeditionary Medal for his actions in helping the wounded. In 1939, he was awarded a Nieman Fellowship at Harvard. P. French, *Through the Looking Glass: China's Foreign Journalists from Opium Wars to Mao* (Hong Kong: Hong Kong University Press, 2009), 205.
36. James, "Hong Kong," 53.
37. J. Keswick, "The Romance of Trade in Hong Kong," in *Hong Kong Business Symposium: A Compilation of Authoritative Views on the Administration, Commerce and Resources of Britain's Far East Outpost*, ed. J. M. Braga (Hong Kong: South China Morning Post, 1957), 38.
38. HKG, *A Problem of People* (Hong Kong: Government Printer, 1957), 28.
39. HKG, Annual Report 1960, 165.
40. HKG, Annual Report 1959, 130.

4. "A PROBLEM OF PEOPLE"

41. HKG, Annual Report 1959, 13, and Annual Report 1964, 106.
42. HKG, Annual Report 1961, 29. The author was most likely Kenneth Barnett. Holdsworth and Munn, *Dictionary of Hong Kong Biography*, 19.
43. HKG, Legislative Council, Report of Proceedings, March 6, 1958.
44. The significant physical presence of the British military in the colony, and CLP's interactions with the military, deserves further research. J. E. Strickland, *Southern District Officer Reports: Islands and Villages in Rural Hong Kong, 1910–60* (Hong Kong: Hong Kong University Press, 2010), 295–96.
45. See K. Ross, *Fast Cars, Clean Bodies: Decolonization and the Reordering of French Culture* (Cambridge, MA: MIT Press, 1996), for an analysis of the imposition and adoption of a modern hygienic future in post-1945 France.
46. Hong Kong Heritage Project (hereafter HKHP, SEK-04-020, C.2-K-5 (letter of November 20, 1948).
47. Albert Raymond and his uncle Abraham Jacob Raymond were on the board of Bombay's Raymond Woolen Mills, part of the E. D. Sassoon empire, which was named after Albert and Abraham's family. When Albert Raymond died at the age of seventy-four, he was a director of Hongkong Land, General Commercial Corp., Hongkong Rope Manufacturing, and Lane Crawford as well as CLP. Lawrence Kadoorie eulogized Raymond as "an unusual man of great knowledge and charm—both student and teacher, a scholar in Hebrew, Latin, and Greek, yet a business executive of unrivalled experience and understanding" (Carl T. Smith collection, No. 57-1045). For Raymond's role in the "buy Jewish" campaign, see Amelia Allsop, *Past & Present* 4 (2017): 6. The 1934 date comes from a recollection by his son (Carl T. Smith collection, No. 57-1046).

    See also https://jhshk.org/community/the-jewish-cemetery/burial-list/raymond-albert/. Abraham Jacob Raymond was one of the shareholders of the Peak Tramway Co. following a shareholding restructuring in 1905, the year in which the Kadoorie family was first associated with the Tramway. D. Waters, "Hong Kong Hongs with Long Histories and British Connections," *Journal of the Hong Kong Branch of the Royal Asiatic Society* 30 (1990): 219–56.
48. Benham was born and educated in England; he went to Shanghai where he was manager of Sassoons. He served in the Royal Air Force in World War II and returned to Shanghai in 1945 before transferring to Hong Kong in 1948. In addition to serving as a CLP director, he was also a director of Hongkong Land, Hongkong Tramways, Peak Tramways, Hongkong & Kowloon Wharf & Godown Co. and the Star Ferry. Carl Smith Collection, No. 36-445, citing *South China Morning Post* (hereafter *SCMP*), August 20, 1962.
49. For Kadoorie's May 1946 account of Jews in Hong Kong, see "Jewish Population of Hong Kong," A Borrowed Place: Jewish Refugees in Hong Kong, 1938–1953, hongkongrefuge.wordpress.com/2016/04/07/a-history-of-jews-in-hong-kong-1946/.
50. Grantham, *Via Ports*, 123–24.
51. HKHP, SEK-04-027, C-2.L.1, part 3. "The Week on Ice House Street," *SCMP*, February 1963. A penciled notation in the HKHP clipping reads "Crew," most likely Leonard Crew, a prominent shareholder and a broker with whom Kadoorie corresponded and who spoke against the merger with HKE at the June 7, 1962 Extraordinary General Meeting (HKHP, SEK-04-024, C.2-K-9/14, 33, 52).
52. F. H. H. King, *The Hongkong Bank in the Period of Development and Nationalism, 1941–1984* (Cambridge: Cambridge University Press, 1991), 684.

4. "A PROBLEM OF PEOPLE"

53. HKHP, C-2-K-3/1; HK PRO, HKRS 163-1-2438; 10/4576/59, "Electricity Supply Companies Commission: Submissions by Government Departments," No. 11.
54. CLP, Annual Report, various years.
55. CLP, Annual Report 1955, 7.
56. CLP, Annual Report 1935.
57. This number presumably counted factories that were large enough to fall under the government regulation requiring factories employing more than fifteen people to register. "Recorded and registered factories" are defined as follows: Factories subject to registration under the Factories and Workshops Ordinance are those that employ twenty or more people or use power-driven machinery; these are subject to inspection. Factories that are "recorded" and thus are "kept under observation" are those that employ fifteen to nineteen workers, those in which women or young workers are employed, or those in which the materials or processes may present health or safety hazards to workers. Szczepanik, *Economic Growth*, 61.
58. Szczepanik contends that the government estimate of 100,000 workers in unregistered establishments or engaged in outwork was too low. Szczepanik, *Economic Growth*, 63.
59. Braga, *Hong Kong Business Symposium*, 36. This reads as if it is Keswick's account wrapped up in the guise of a visitor to the colony.
60. M. Keswick, *The Thistle and the Jade: A Celebration of 150 Years of Jardine, Matheson & Co.* (London: Octopus Books, 1982), 262–65.
61. HKHP, R.2-A-1/28.
62. Author interview, Michael Kadoorie, December 2016.
63. HKHP (2011), 22; S. L. Wong, *Emigrant Entrepreneurs: Shanghai Industrialists in Hong Kong* (Hong Kong: Oxford University Press, 1988), 48.
64. Nanyang Holdings (2011), "Proposals Relating to a General Mandate for Repurchase by the Company of its Own Shares and Re-election of Retiring Directors," http://www.nanyangholdingslimited.com/file/circular_listing/eng/E_Cir_20110413.pdf.
65. Rong founded the China International Trust and Investment Co. (CITIC) in 1979, earning him the sobriquet "red capitalist." In 1987, his son, Larry Yung Chi-kin (Rong Zhijian), founded the predecessor to the group's Hong Kong–listed company, Citic Pacific, and was chairman until he resigned in 2009 after police investigations into large foreign exchange losses.
66. Author interview, Michael Kadoorie, December 2016.
67. Szczepanik, *Economic Growth*, 107.
68. S. Tsang, *A Modern History of Hong Kong* (Hong Kong: Hong Kong University Press, 2004), 163.
69. T. W. Ngo, *Hong Kong's History: State and Society Under Colonial Rule* (London: Routledge, 1999), 94.
70. Quoted in Wong, *Emigrant Entrepreneurs*, 24–25.
71. Grantham, *Via Ports*, 155.
72. J. M. Carroll, *Edge of Empires: Chinese Elites and British Colonials in Hong Kong* (Cambridge, MA: Harvard University Press, 2005), 57, 67. Both Kadoorie's role in wooing the Shanghainese and the possibility that they would go elsewhere should be critically examined. Wong suggests that Hong Kong and Taiwan were the only practical alternatives for many Chinese businesses (*Emigrant Entrepreneurs*,

4. "A PROBLEM OF PEOPLE"

20–25). Somewhat contradictorily, Wong also notes the sense of insecurity that prevailed in Hong Kong in the years immediately after 1949, with one textile executive considering establishing operations in Argentina, saying "Everyone in Hong Kong was scared that the Communists would be coming. We had to have a place to go" (38–39).

73. CLP, Annual Report 1952.
74. CLP, Annual Report 1953; see also *Past & Present* 1 (2013).
75. "An Electronic Brain for China Light by Next October," *China Mail*, November 10, 1962; *Past & Present* 2 (2014): 2–3.
76. CLP's service area included substantially all of the territory except Hong Kong Island and the smaller Lamma Island, or about 1,025 square kilometers. https://www.yearbook.gov.hk/2016/en/pdf/Facts.pdf.
77. "1950s Hong Kong: Rise of Industry and Tourism," *Past & Present* 1 (2013): 2.
78. CLP, Annual Report 1953, 9.
79. CLP, Annual Report 1954, 288–89.
80. CLP, Annual Report 1955, 7.
81. CLP, Annual Report 1953, 5.
82. CLP, Annual Report 1957, 8.
83. HKHP (2011), 30.
84. CLP, Annual Report 1960, 10–11.
85. HKG, Annual Report 1957, 292.
86. E. Taus, "Electrical Appliances," in Braga, *Hong Kong Business Symposium*, 426.
87. W. S. B. Wong, "Air Conditioning," in Braga, *Hong Kong Business Symposium*, 388–89.
88. American Film Distributors, "Film Distribution," in Braga, *Hong Kong Business Symposium*, 427.
89. Kadoorie responded to a proposal from James H. Moore of Mei Foo Investments in 1966 suggesting that CLP finance air conditioner purchases in a thirty-five-hectare Lai Chi Kok apartment complex that the company was constructing: "The CL&P is certainly interested in expanding sale of current and your proposal that they finance air conditioning units for your new development at Laichikok will be looked at carefully" (HKHP, SEK-4-002, C.2-9/120-1).
90. "The Hongkong & Shanghai Hotels, Limited," in Braga, *Hong Kong Business Symposium*, 445. For HSBC branches, see Y. C. Jao, "Financing Hong Kong's Early Postwar Industrialization: The Role of the Hongkong and Shanghai Banking Corporation," in *Eastern Banking: Essays in the History of the Hongkong and Shanghai Banking Corporation*, ed. F. H. King (London: Athlone Press, 1983), 553–54.
91. King, *Hongkong Bank*, 369.
92. N. Cameron, *Power: The Story of China Light* (Hong Kong: Oxford University Press, 1982) opp. 124.
93. CLP, Annual Report 1960, 30.
94. CLP, Annual Report 1956, 6.
95. CLP, Annual Report 1956, 6–7.
96. CLP, Annual Report 1957, 8–9.
97. CLP, Annual Report 1958, 12.
98. *Far Eastern Economic Review* 27, no. 9 (1959): 317–22.
99. *Far Eastern Economic Review* 36 (1996): xi–xii.

100. CLP, *Submissions*, 6.
101. "Cheaper Electricity and More Water," *SCMP*, August 28, 1959.
102. CLP, Annual Report 1960, 15–16. The Calder reactors were the so-called Magnox variety and had the dual goal of producing plutonium (for weapons) and electricity; their efficiency in producing electricity was relatively poor.
103. HKHP 201/117.
104. CLP, Annual Report 1981, 6, and Annual Report 1982, 11.
105. HKHP, CLP-3A-01 (memo from Laufer on nuclear power, March 22, 1977).

5. ELECTRICITY AS A POLITICAL PROJECT

1. For an overview of how governance issues were conceptualized globally in this period, see M. Bevir, *Governance: A Very Short Introduction* (Oxford: Oxford University Press, 2012).
2. T. Hughes, *Networks of Power: Electrification in Western Society, 1880–1930* (Baltimore, MD: Johns Hopkins University Press, 1983); L. Hannah, *British Electricity Before Nationalisation: A Study of the Development of the Electricity Supply Industry in Britain to 1948* (London: Macmillan, 1979).
3. W. J. Hausman, P. Hertner, and M. Wilkins, *Global Electrification: Multinational Enterprise and International Finance in the History of Light and Power, 1878–2007* (Cambridge: Cambridge University Press, 2008).
4. Hong Kong Heritage Project (hereafter HKHP), SEK-04-025, C-2-K-10 (adjacent to /122). Emphasis added.
5. For treatments of this subject, particularly in the United States, see T. Hughes, *Networks of Power*; J. Lambert, *The Power Brokers: The Struggle to Shape and Control the Electric Power Industry* (Cambridge, MA: MIT Press, 2015); and F. McDonald, *Insull* (Chicago: University of Chicago Press, 1962).
6. Some of the "water" references refer to navigable waters, but hundreds refer to the water supply system.
7. L. Kadoorie, "N. T. Agreement," *South China Morning Post* (hereafter *SCMP*), July 11, 1962.
8. HKHP, SEK-04-025, C.2-K-10/26.
9. HKHP, SEK-04-026, C.2-K-11/E7.
10. J. D. Poulter, *An Early History of Electricity Supply: The Story of the Electric Light in Victorian Leeds* (London: P. Peregrinus, 1986).
11. Hughes, *Networks of Power*; Lambert, *The Power Brokers*.
12. McDonald, *Insull*, 91.
13. McDonald, *Insull*, 113–14; Lambert, *The Power Brokers*, 15–17.
14. Lambert, *The Power Brokers*, 51–92.
15. R. A. Caro, *The Path to Power* (New York: Knopf, 1982), 326–328, 516–28.
16. S. S. Kale, *Electrifying India: Regional Political Economies of Development* (Stanford, CA: Stanford University Press, 2014), 24–53.
17. Hong Kong Public Records Office (hereafter HK PRO), 9/4576/59, HKRS No. 41 D-S No. 1-9815-1, Nos. 22–23.
18. The parallel between the democratization of electricity and the democratization of information could be explored. The early movements against private cable

monopolies also called for a penny post—allowing letters to be mailed anywhere within the Empire for only a penny—as a way of democratizing communications. D. Winseck and R. Pike, *Communication and Empire: Media, Markets, and Globalization, 1860–1930* (Durham, NC: Duke University Press, 2007), 142–43.

19. E. F. Szczepanik, *The Economic Growth of Hong Kong* (London: Oxford University Press, 1958), 85–87.
20. The 1976 government *Report of the Working Party of Electrification of Squatter Areas* found that eighty percent of residents in Tai Yuen village obtained electricity through illegal tapping; others used kerosene lamps and stoves. Private generators, regarded as unsafe by the government, also supplied a good deal of electricity in squatter areas. Hong Kong Government, *Report of the Working Party of Electrification of Squatter Areas* (Hong Kong: Government Printer, 1976), 7.
21. J. C. Scott, *Seeing Like a State: How Certain Schemes to Improve the Human Condition Have Failed* (New Haven, CT: Yale University Press, 1998).
22. HKHP, SEK-04-016, C.2-K-1/ Letter dated April 13, 1956.
23. HKHP, SEK-04-017, C.2-K-2.
24. Szczepanik, *The Economic Growth of Hong Kong*, 105.
25. A. Coates, *A Mountain of Light: The Story of the Hongkong Electric Company* (London: Heinemann, 1977), 200–202.
26. "Hongkong's Power Problems," *Hongkong Tiger Standard* (hereafter *HKTS*), November 30, 1958.
27. Articles mentioned include: "Actions Speak Louder," *HKTS*, December 5, 1958; "Power Breaks Halt Operations of HK Plants," *HKTS*, December 9, 1958; "Civic Leaders, Industrialists Express Concern Over Power Breaks," *HKTS*, December 10, 1958.
28. D. E. Nye discusses changing expectations about electricity availability in *When the Lights Went Out: A History of Blackouts in America* (Cambridge, MA: MIT Press, 2010).
29. Alan Castro, "Tiger Roars for HK," *HKTS*, March 26, 1999.
30. J. A. Lent, *The Asian Newspapers' Reluctant Revolution* (Ames: Iowa State University Press, 1971), 15, 63.
31. I was the publisher and editor-in-chief at *The Standard* in 2004–2006 and editor-in-chief at the *South China Morning Post* in 2006–2007.
32. The surcharge had long been contentious. The HKE 1956 annual report notes a representation about the fuel surcharge by various bodies, including the Chinese Manufacturers Union and the Chinese General Chamber of Commerce (HKE, Annual Report 1956, 5).
33. "The Surcharge," *China Mail*, undated article in HKHP, SEK, 04-027, C.2-L-1 (adjacent to /6).
34. "Development Finance Corp. Urged," *SCMP*, September 30, 1959; see also N. Monnery, *Architect of Prosperity: Sir John Cowperthwaite and the Making of Hong Kong* (London: London Publishing Partners, 2017), 124–25; and S. L. Wong, *Emigrant Entrepreneurs: Shanghai Industrialists in Hong Kong* (Hong Kong: Oxford University Press, 1988), 89–90.
35. One of the most prominent Chinese people in the colony during this period, M. K. Lo, was married to former CLP director Robert Ho Tung's daughter; Lo's father, Eurasian comprador Lo Cheung-shiu, had earlier served on the CLP

5. ELECTRICITY AS A POLITICAL PROJECT

board. In 1935 M. K. Lo replaced Robert Kotewall on the Legislative Council, "where he soon developed a reputation for being the most vocal critic of the colonial government." John Carroll quoted in M. Holdsworth and C. Munn, *Dictionary of Hong Kong Biography* (Hong Kong: Hong Kong University Press, 2012), 279. Lo was also the chairman of the Anti-Tuberculosis Association, set up by Dr. Selwyn Selwyn-Clarke in the late 1930s. M. Jones, "Tuberculosis, Housing and the Colonial State: Hong Kong, 1900–1950," *Modern Asian Studies* 37, no. 3 (2003): 671.

36. HKHP, SEK-04-024, C.2-K-1/7-12.
37. L. F. Goodstadt, *Uneasy Partners: The Conflict Between Public Interest and Private Profit in Hong Kong* (Hong Kong: Hong Kong University Press, 2005), 176–77.
38. Author interview, Hilton Cheong-leen, 2016.
39. Holdsworth and Munn, *Dictionary of Hong Kong Biography*, 28.
40. Personal communications with Hilton Cheong-leen, November and December 2017. See also H. Cheong-leen, *Hongkong Tomorrow: A Collection of Speeches and Articles by Hilton Cheong-leen* (Hong Kong: Local Property & Printing, 1962), and V. Maher, "Hilton Cheong-leen," *SCMP*, July 28, 2003.
41. HKHP, SEK-04-27, C.2-L-1/14.
42. "HK Electric to Help Govt Inquiry," *China Mail*, April 10, 1959.
43. "Hongkong's Power Supplies," *SCMP*, April 11, 1959.
44. The Central Electricity Board was the prenationalization entity; it was replaced in 1948 by the Central Electricity Authority and in 1957 by the Central Electricity Generating Board. Dieter Helm, *Energy, the State, and the Market: British Energy Policy Since 1979* (Oxford: Oxford University Press, 2003), 18, 19. Kadoorie did not appear to be distinguishing between the different phases of this organization.
45. HKHP, SEK-04-048, C.2-U-1/97.
46. HKHP, SEK-04-024, C.2-K-1/21, 25.
47. "Enquiry Into Position of Electricity Companies," *SCMP*, June 11, 1959.
48. "Electricity Commission Chairman," *SCMP*, August 24, 1959.
49. HK PRO, HKRS 41-1-9813; "Financial arrangements"; Floating Sheet 1 and HK PRO HKRS 41-1-9814 secretariat GR6/4576/5, "Electricity Supply Companies Commission; Financial Arrangement."
50. HK PRO, 9/4576/59; HKRS No. 41 D-S No. 1-9815-1, "Electricity Supply Companies Commission: Miscellaneous Correspondence," Nos. 1, 4, 7.
51. HK PRO, 9/4576/59; HKRS No. 41 D-S No. 1-9815-1, "Electricity Supply Companies Commission: Miscellaneous Correspondence," Nos. 10, 12. No record of the meeting between Whitelegge and Barber has been found. The Carl T. Smith card collection records Charles Hill Barber as an "American merchant" who lived at 8A Ho Koi Road and died January 3, 1973 (Carl T. Smith collection, No. 35–864).
52. "Electricity Commission Chairman," *SCMP*, August 24, 1959.
53. HK PRO, HKRS No. 163 / D-S No. 1-2438, 10/4576/59, "Electricity Supply Companies Commission: Submissions by Government Departments," No. 19.
54. HK PRO, "Electricity Supply Companies Commission." Holmes attached a number of other documents, notably the No. 15 series.
55. HK PRO, "Electricity Supply Companies Commission: Submissions by Government Departments," No. 11.
56. "Fuel Clause Explained," *China Mail*, October 23, 1959. The surcharge was advertised in the press August 11, 1951. China Light & Power, *Submissions and Final*

5. ELECTRICITY AS A POLITICAL PROJECT

Address of the China Light & Power Co., Ltd., to the Electricity Supply Companies Commission 1959 (Hong Kong: CLP, 1959), 32, 33, 34; HKE, Annual Report 1951.
57. HKHP, SEK-04-017, C.2-K-2/14.
58. Most of the electricity debate took place in an English-language setting. Letters of any sort were uncommon in Hong Kong's Chinese-language newspapers.
59. HKG, Annual Report 1959: 235.
60. Szczepanik, *The Economic Growth of Hong Kong*, 89–90.
61. "Nationalisation Principle Wrong," *HKTS*, August 6, 1962, quoting shareholder Dr. D. Engel.
62. "Integration of Two Electric Companies," *SCMP*, November 17, 1959.
63. "Objections to Control Reiterated," *SCMP*, November 18, 1959.
64. "Hong Kong Conditions," *SCMP*, January 21, 1960.
65. E. Elliot [Tu], *Colonial Hong Kong in the Eyes of Elsie Tu* (Hong Kong: Hong Kong University Press, 2003).
66. HK PRO, HKRS 163-1-2271 CR1/4576/58, "Correspondence from Mr. W. S. Edwards Concerning the Electricity Supply Companies Commission."
67. HKE does not have employee records from this period. Personal communication, Alex Ng, 2018.
68. HKHP, SEK-04-017, C.2-K-2/31: 3.
69. "Representations on Electricity," *SCMP*, August 8, 1959. The original article has no start to this quote.
70. "Complaints Against H. K. Power Companies," *SCMP*, August 21, 1959.
71. HK PRO, HKRS No. 163. D-S No. 1-2438, 10/4576/59, "Electricity Supply Companies Commission: Submissions by Government Departments," No. 15-1; No. 15-3, No. 15-4.
72. HK PRO, HKRS 163-1-2438, 10/4576/59, "Electricity Supply Companies Commission: Submissions by Government Departments," No. 15-5, Enclosure III, "List of those New Territories Villages which have a supply of Electricity." Islands were not included. Nor were the market towns, all of which were electrified.
73. Lambert, *The Power Brokers*, 22–28, 51–92; Caro, *The Path to Power*, 516–28.
74. Concern about the theft of electricity is a recurring theme that goes back at least to the late 1920s. An agenda item for the June 30, 1927, Board of Directors meeting read: "Work manager's report on theft of current and prosecution of Chinese hotels."
75. HK PRO, HKRS 163-1-2438, 10/4576/59, "Electricity Supply Companies Commission: Submissions by Government Departments," No. 19-3, No. 19-4.
76. HK PRO, HKRS 163-1-2438, 10/4576/59, "Electricity Supply Companies Commission: Submissions by Government Departments," No. 15-4.
77. HK PRO, HKRS 163-1-2438; 10/4576/59, "Electricity Supply Companies Commission: Submissions by Government Departments," No. 13.
78. HK PRO, HKRS 163-1-2438; 10/4576/59, "Electricity Supply Companies Commission: Submissions by Government Departments," No. 12.
79. HK PRO, HKRS 163-1-2438; 10/4576/59, "Electricity Supply Companies Commission: Submissions by Government Departments," Nos. 9, 9-1.
80. HK PRO, HKRS 163-1-2438; 10/4576/59, "Electricity Supply Companies Commission: Submissions by Government Departments," No. 8-3.

5. ELECTRICITY AS A POLITICAL PROJECT

81. HK PRO, HKRS 163-1-2438; 10/4576/59, "Electricity Supply Companies Commission: Submissions by Government Departments," No. 14.
82. HK PRO, HKRS 163-1-2438; 10/4576/59, "Electricity Supply Companies Commission: Submissions by Government Departments," No. 16.
83. HKHP, SEK-04-019, C.2-K-4/37.
84. HKHP, SEK-04-019, C.2-K-4/37: 441.
85. HKHP, SEK-04-019, C.2-K-4/37: 442.
86. HKHP, SEK-04-48, C.-2.U.1/14.
87. On January 20, 1960, the day that the ESCC's report came out, Acting Colonial Secretary Edmund Teesdale wrote Kadoorie that the government wanted to ensure the "safeguarding of assets" in preparation for the imposition of controls on the electricity supply companies. Dividends were frozen at a rate no higher than those paid April 10, 1959, which meant a cut for CLP shareholders. Directors would be liable for any excess payments (HKHP, SEK-04-018, C.2-K-3/3). On January 30, 1960, Colonial Secretary Claude Burgess wrote CLP that the Governor was "unable to comply with your request" to pay the same dividend as the previous year (HKHP, SEK-04-018, C.2.K.3/11). Cowperthwaite also refused to allow an increased dividend. He vowed that he was "not going to be placed in a position where both companies would work up the dividend at the cost of Government and the consumer" (HKHP, SEK-04-025, C.2-K-10/18).
88. "Govt. Statement on Electricity Report," *SCMP*, March 23, 1960.
89. "Electric Companies' Proposals Rejected," *SCMP*, January 28, 1961.
90. "Wanted: Better Deal," *HKTS*, June 8, 1962; "China Light Co. Meeting," *SCMP*, June 8, 1962; "Extraordinary General Meeting of China Light and Power Co., Ltd.," *SCMP*, June 9, 1962; "Opposition to Merger Terms," *SCMP*, June 9, 1962.
91. "Light on the China Light Company," *HKTS*, April 26, 1959.
92. "Hongkong's Power Supplies," *SCMP*, April 11, 1959.
93. HKHP, SEK-04-025, C.2.-K.-10/18 and SEK-04-023 C.2-K-8/56.
94. HKHP, SEK-04-018, C.2-K-3/38.
95. HKE, Annual Report 1960: 6.
96. "Monopoly Profits Denied," *China Mail*, October 22, 1959.
97. HKHP, SEK-04-023, C.2-K-8, especially /1, 20, 31, 36, 48, 49, 54. For merger announcement, see "Light Charges Will Fall," *HKTS*, May 2, 1962; "H. K.'s Power Companies to Be Merged" and "Electricity Undertakings," *SCMP*, May 2, 1962.
98. HKHP, SEK-04-023, C.2-K-8/1.
99. HKHP, SEK-04-023, C.2-K-8/1.
100. HSBC Manager Michael Turner surmised that it was Finance Secretary Arthur Clarke's unhappiness with rights offerings that led him to propose nationalization. In an April 30, 1956, letter to Kadoorie, commenting on Clarke's letter, Turner noted Clarke's goal of "putting an end to the practice of issuing rights to new shares on such generous terms as are being done now" (HKHP, SEK-04-016, C.2-K-1).
101. HKHP, SEK-04-024, C.2-K-9/ 31.
102. HKHP, SEK-04-024, C.2-K-9/31 and SEK-04-025, C.2-K-10/27, 28. For Cowperthwaite's proposal that the government take shares, see HKHP, SEK-04-025, C.2-K-10/35.

5. ELECTRICITY AS A POLITICAL PROJECT

103. HKHP, SEK-04-025, C.2-K-10/18.
104. HKHP, C.2-K-10/27.
105. HKHP, SEK-04-023, C.2-K-8/13.
106. HKHP, SEK-04-023, C.2-K-8/12.
107. HKHP, SEK-04-019, C.2-K-4/37.
108. HKHP, SEK-04-020, C.2-K-5/7.
109. HKHP, SEK-04-023, C.2-K-8/13.
110. HKHP, SEK-04-023, C.2-K-8/21.
111. McDonald, *Insull*, 71–73.
112. CLP, Annual Report, "Statement of accounts for the period from January 1, 1942 until August 31, 1945," lists an overdraft of $742,362 from HSBC. This is the only overdraft entry found in the CLP accounts. Between 1949 and 1955, bank borrowing accounted for 3–12 percent of CLP's financing, an extremely low figure for an electricity supply company (CLP, *Submissions and Final Address*, 4). Retentions, including depreciation, accounted for 63–82 percent.
113. "Extraordinary General Meeting of China Light and Power Co., Ltd.," *SCMP*, June 9, 1962.
114. The Kadoories held 38.93 percent at the time of the ESCC hearings and there was only minor, if any, variation by the time of the 1962 shareholders' meeting (HKHP, SEK-04-016, C.2-K-1).
115. HKHP, SEK-04-025, C.2-K-10/11.
116. HKHP, SEK-04-025, C.2-K-10/18.

6. "DIE-HARD REACTIONARY" IN THE EXPANDING COLONIAL STATE

1. R. Chernow, *Titan: The Life of John D Rockefeller, Sr.* (New York: Vintage Books, 2004), 557.
2. A. Wright and H. A. Cartwright, *Twentieth Century Impressions of Hongkong, Shanghai, and Other Treaty Ports of China: Their History, People, Commerce, Industries, and Resources* (London: Lloyd's Greater Britain Publishing, 1908), 213.
3. Hong Kong Heritage Project (hereafter HKHP), SEK-4-028, C.2-L-2. Hong Kong Government (hereafter HKG), "China Light and Power Co. Ltd.," *Daily Information Bulletin Supplement*, November 23, 1964.
4. Author interview, December 2016.
5. HKHP, SEK-04-029, Part 1, C.2-0-1/93.
6. HKHP, SEK-04-029, Part 1, C.2-0-1/1, 93.
7. HKHP, SEK-04-029, Part 1, C.2-0-1/3.
8. HKHP, SEK-04-029, Part 1, C.2-0-1/1.
9. L. Kadoorie, "Electricity Success Story," *Financial Times*, July 19, 1965.
10. J. M. Carroll, *A Concise History of Hong Kong* (Hong Kong: Hong Kong University Press, 2007), 144.
11. P. Roberts and J. Carroll, eds., *Hong Kong in the Cold War* (Hong Kong: Hong Kong University Press, 2016), 44.

6. "DIE-HARD REACTIONARY" IN THE COLONIAL STATE

12. HKHP, SEK-04-029, Part 1, C.2-0-1/68, 92.
13. HKHP, SEK-4A-015: 21. A Freedom of Information Act request filed with the U.S. government elicited no further information.
14. N. Cameron, *Power: The Story of China Light* (Hong Kong: Oxford University Press, 1982), 201.
15. HKHP, SEK-04-031, C.2-0-3/90.
16. HKHP, SEK-04-A-015, 23.
17. China Light & Power Co. (CLP), *Golden Memories of Tsing Yi Power Station* (Hong Kong: CLP, 1999), 5.
18. Kadoorie, "Electricity Success Story."
19. HKG, Annual Report 1956, 68–69.
20. L. Kadoorie, "China Light & Power Co. Ltd.: 1901–1918" (Hong Kong: CLP, 1977).
21. CLP, Annual Report 1925, noted the payment of an extra bonus to staff for overtime work during the strike of $1,450, in addition to staff bonuses of $8,866.
22. HKHP, CLP-03A-09 (October 19, 1979 interview). The strike lasted thirty-four days, and "military and trades union officials from Canton" were said to have supported strikers who cut cables, broke into substations, and damaged relays. The strike broke out almost as soon as CLP took over control of the company from the military and was settled July 26, 1946.
23. A. Allsop, ed., *St. George's Building: A Brief Portrait* (Hong Kong: Hong Kong Heritage Project, 2016), 97; HKHP, SEK-1A-055: 22.
24. CLP, *Golden Memories*, 25.
25. Cameron, *Power*, 215.
26. HKHP, SEK-04-003, Part 2, C.2-A-10/166.
27. HKHP, SEK-04-003, Part 2, C-2.A-10/123.
28. HKHP, SEK-04-003, Part 2, C-2.A-10/129, 108a.
29. HKHP, SEK-04-001, Part 1 of 2, C.2-A-8/1, 48; SEK-04-048, C.2-U-1/ 75, 97; SEK-04-17, C.2-K-2/44, 45.
30. "City Escalator in Use," *SCMP*, August 2, 1957.
31. C. J. Ure, "Elevators," in *Hong Kong Business Symposium: A Compilation of Authoritative Views on the Administration, Commerce and Resources*, ed. J. M. Braga (Hong Kong: South China Morning Post, 1957), 46.
32. C. F. Wood, "China Light & Power Co., Ltd.," in Braga, *Hong Kong Business Symposium*, 312–15.
33. CLP, Annual Report 1957, 8–9.
34. CLP, Annual Report 1957, 9.
35. CLP, Annual Report 1960, 12.
36. Kadoorie told the ESCC that Hong Kong suffered from inadequate telephones, buses, piers, ferry services, and hospitals: "Too little, too late, and at too high a cost." CLP, by contrast, had built at a low cost "one of the finest power stations, for its size, to be found anywhere." CLP, *Submissions and Final Address of the China Light & Power Co., Ltd., to the Electricity Supply Companies Commission 1959* (Hong Kong: CLP, 1959), 46.
37. *Past & Present* 2 (2014), 2.
38. CLP, Annual Report 1953, 9; *Past & Present* 1 (2013), 2.

6. "DIE-HARD REACTIONARY" IN THE COLONIAL STATE

39. HKHP, SEK-04-003, C.2-A-1-0/16, 19; HKG, Annual Report 1953, 257.
40. CLP, Annual Report 1960, 11.
41. CLP, Annual Report 1957, 8; *Past & Present* 1 (2013), 2.
42. E. F. Szczepanik, *The Economic Growth of Hong Kong* (London: Oxford University Press, 1958), 69. See also http://www.polyu.edu.hk/web/en/about_polyu/facts_figures_development/history/index.html.
43. CLP, Annual Report 1960, 15.
44. CLP, Annual Report 1948.
45. CLP, Annual Report 1953, 5.
46. C. K. Lamb, "Supply Companies Ahead of Hong Kong Industry," *Asian Industry*, April 1966, 12.
47. *Past & Present* 2 (2014), 2–3.
48. HKG, *Administrative Reports, 1940–41*, 94.
49. CLP, Annual Report 1954, 8; CLP, Annual Report 1956, 5.
50. Wright and Cartwright, *Twentieth Century Impressions*, 240. Waters writes that Jardine, Matheson imported the first ice in 1843 and that the Ice House Co. was established in 1845. Ice-making began in 1874 at East Point. D. Waters, "Hong Kong Hongs with Long Histories and British Connections," *Journal of the Hong Kong Branch of the Royal Asiatic Society* 30 (1990): 29.
51. R. Hutcheon, *Wharf: The First Hundred Years* (Hong Kong: Wharf Holdings, 1986), 178.
52. Hutcheon, *Wharf*, 188.
53. CLP, Annual Report 1907.
54. Clark, A. D., ed., *100 Years of Electricity* (Hong Kong: Hong Kong Electric Co., 1990), 26.
55. C. Stratford, "New Kowloon Power Station of the China Light and Power Co. (1918) Ltd.," *Far Eastern Review*, June 1931, 371.
56. CLP, Annual Report 1938.
57. HKHP, CLP-3A-09.
58. Szczepanik, *Economic Growth*, 66.
59. Szczepanik, *Economic Growth*, 93.
60. R. Hughes, *Borrowed Place, Borrowed Time: Hong Kong and Its Many Faces* (London: Andre Deutsch, 1976), 13. There were other mining activities in the colony in addition to extensive sand and granite quarrying. In 1956 iron ore was exported to Japan, lead ore to the United Kingdom and Europe, and tungsten (known in Hong Kong by its alternate name, wolfram) and graphite to the United Kingdom and the United States (HKG, *Annual Report 1956*, 107). Some 3000 illegal wolframite miners worked the hills above Sha Lo Wan in the early 1950s; see J. E. Strickland, *Southern District Officer Reports: Islands and Villages in Rural Hong Kong, 1910–60* (Hong Kong: Hong Kong University Press, 2010), 66–67. CLP supplied electricity to an iron ore mine on Ma On Shan that employed "almost entirely immigrant labor from North China" (CLP, Annual Report 1954, 7, 20; HKG, Annual Report 1956, 57).
61. Cameron, *Power*, 242–43.
62. CLP, *Golden Memories*, 41; CLP, Annual Report 1981, 13.
63. HKG, Annual Report 1982, 164.
64. HKHP, C.2-AA1/C. China Project.

6. "DIE-HARD REACTIONARY" IN THE COLONIAL STATE

65. The pivotal Third Plenum, which is generally regarded as the beginning of the reform and opening policy, took place in December 1978. Kadoorie's visit is a reminder that change was already underway.
66. CLP, *Power for Prosperity: An Overview of China Light & Power and its Role in Hong Kong's Development* (Hong Kong: CLP, 1991), 5.
67. CLP, Annual Report 1981, 6.
68. HKG, Legislative Council Secretariat, "China Light and Power Black Point Project: The Government's Monitoring of Electricity Supply Companies" (1999), 35.
69. HKHP, SEK-04-017, C.2-K-2.
70. HKG, Annual Report 1964, 241.
71. "Nationalization Principle Wrong," *HKTS*, June 8, 1962.
72. CLP, Annual Report 1955, 4; Annual Report 1957, 4, 7; Annual Report 1958, 3–5; S. T. Kwok, *A Century of Light* (Hong Kong: CLP, 2001), 75–76.
73. L. Kadoorie, "Chairman's Speech for Inaugural Ceremony of New Power Station" (Hong Kong: CLP, 1940), 2.
74. Kwok, *A Century of Light*, 98.
75. HKE, Annual Report 1974.
76. CLP, Annual Report 1982, 7.
77. HKE, 1990: 73.
78. CLP, Annual Report 1935.
79. CLP, Annual Report 1953, 3–4.
80. HKE, Annual Report 1973, 6–7.
81. M. S. Witkovsky, C. S. Eliel, and K. P. B. Vail, eds., *Moholy-Nagy: Future Present* (Chicago: The Art Institute of Chicago, 2016), 17, 18, 30–31, 145–57, 158.
82. J. F. Wasik, *The Merchant of Power: Sam Insull, Thomas Edison, and the Creation of the Modern Metropolis* (New York: Palgrave Macmillan, 2006), 93.
83. Kwok, *A Century of Light*, 76.
84. CLP, Annual Report 1958, 6.
85. CLP, Annual Report 1960, 30.
86. CLP Annual Report 1959, 3–4.
87. HKG, *Report of the Working Party of Electrification of Squatter Areas* (Hong Kong: Government Printer, 1976). (Hereafter *Report of the Working Party*.)
88. In the 1950s CLP had been unable to get government approval to supply electricity to illegal structures (CLP, 1959: 27).
89. HKG, *Report of the Working Party*, 2–3.
90. HKG, *Report of the Working Party*, 1.
91. HKG, *Report of the Working Party*, 32–36.
92. HKG, *Report of the Working Party*, 4–5, 19.
93. Consciously or not, this location echoed the television series *Below the Lion Rock*. The series premiered in 1972 and looked back nostalgically on Hong Kong's hustling, pragmatic spirit of the 1950s and 1960s.
94. HKG, *Report of the Working Party*, 13–14.
95. HKG, *Report of the Working Party*, 25–30.
96. HKG, *Report of the Working Party*, 14–24, 29.
97. HKG, *Report of the Working Party*, 26–27.
98. HKG, *Report of the Working Party*, 16.
99. HKG, *Report of the Working Party*, 31.

7. "INTELLIGENT ANTICIPATION" FOR "1997 AND ALL THAT"

1. Margaret Thatcher Archive (TA), PREM19-0792, f307; E. F. Vogel, *Deng Xiaoping and the Transformation of China* (Cambridge, MA: Harvard University Press, 2011), 493–97.
2. Hong Kong Heritage Project (HKHP), CLP-3A-02.
3. N. Cameron, *Power: The Story of China Light* (Hong Kong: Oxford University Press, 1982), 256, 258.
4. L. Kadoorie, "1997 and All That" (unpublished manuscript, 1973) 101.1.
5. W. C. Sellar, R. J. Yeatman, and J. Reynolds, *1066 and All That: A Memorable History of England, Comprising All the Parts You Can Remember, Including 103 Good Things, 5 Bad Kings and 2 Genuine Dates* (London: Methuen, 1951), 17.
6. Kadoorie, "1997 and All That" (1973), 401.13.
7. Kadoorie, "1997 and All That" (1973), 201.26.
8. L. Kadoorie, "1997 and All That—Further Thoughts" (unpublished manuscript, 1975), 2.
9. A. W. Grantham, *Via Ports, from Hong Kong to Hong Kong* (Hong Kong: Hong Kong University Press, 1965), 172.
10. E. F. Szczepanik, *The Embargo Effect on China's Trade with Hong Kong* (Hong Kong: Hong Kong University Press, 1958), 85–87.
11. M. Wai-chu Tam, *The Basic Law and Hong Kong: The 15th Anniversary of Reunification with the Motherland* (Hong Kong: Working Group on Overseas Community of the Basic Law Promotion Steering Committee, 2012), 4.
12. Tam, *The Basic Law*, 5.
13. Tam, *The Basic Law*, 8.
14. Tam, *The Basic Law*, 8. This account refers to Heath as the prime minister, although he had left office two months earlier and was then head of the opposition. It is unclear whether the error in Heath's title, which is in the text as well as in a photo caption, was simply an oversight or was an attempt to represent the meeting with Mao as one in which Heath was representing an official British position. The version given of the meeting is contradicted by Heath's account when he met Deng Xiaoping April 6, 1982, and noted that he had first raised the issue of Hong Kong with Mao and Zhou Enlai in 1977. At the 1982 meeting Deng outlined the PRC's proposals for an SAR (TA, PREM-19-0792 f3).
15. www.china.org.cn; this cites the *PLA Daily* of May 25, 1974, as the original source.
16. Tam, *The Basic Law*, 8–9; Vogel, *Deng Xiaoping*, 491.
17. Kadoorie, "1997 and All That" (1973), 401.11–12.
18. Kadoorie, "1997 and All That—Further Thoughts" (1975), 4.
19. Kadoorie, "1997 and All That" (1973), 401.13.
20. TA, PREM19-0788.
21. Kadoorie, "1997 and All That—Further Thoughts" (1975), 4.
22. L. Kadoorie, "1997 and All That" (unpublished manuscript, 1976), 4.
23. Kadoorie, "1997 and All That" (1976), 4.
24. HKHP, CLP-3A-01, C.2-D-1.
25. HKHP, SEK-4A-015, 21, 23.

26. L. Kadoorie, "The Kadoorie Memoir," in D. Leventhal and L. Kadoorie, *Sino-Judaic Studies: Whence and Whither: An Essay and Bibliography and the Kadoorie Memoir* (Hong Kong: Hong Kong Jewish Chronicle, 1985), 89.
27. L. Kadoorie, "The Kadoorie Memoir," 83–88.
28. HKHP, CLP-3A-02.
29. Kadoorie, "1997 and All That" (1973), 501.32, 501.33.
30. Kadoorie, "1997 and All That—Further Thoughts" (1975), 12.
31. Kadoorie, "1997 and All That" (1973), 501.42; Kadoorie, "1997 and All That—Further Thoughts" (1975), 10–11.
32. Kadoorie, "1997 and All That" (1973), 501.45.
33. Kadoorie, "1997 and All That—Further Thoughts" (1975), 1, 5.
34. Kadoorie, "1997 and All That" (1976), 5.
35. Kadoorie, "1997 and All That" (1976), 73; Westphal letter dated November 9, 1973.
36. TA, PREM19-0788, September 28, 1982, "Speech Opening Hong Kong Power Station (Castle Peak A)."
37. Kadoorie, "1997 and All That" (1973), 501.45.

8. SING THE CITY ELECTRIC

1. Y. C. Jao, *Banking and Currency in Hong Kong: A Study of Postwar Financial Development* (London: Macmillan, 1974), 5. The Hong Kong government did not collect statistics on overall economic growth, so these are estimates, but there is little doubt that Hong Kong's economy was one of the world's fastest-growing.
2. The issue was raised at a meeting between Zhao Ziyang and the Lord Privy Seal on January 6, 1982. Margaret Thatcher Archive, PREM19-0792 f299.
3. The notion of hybridity follows Osterhammel and Petersson, who define it as "new cultural elements being creatively adapted to mesh with existing ones." J. Osterhammel and N. P. Petersson, *Globalization: A Short History* (Princeton, NJ: Princeton University Press, 2005), 7.
4. Hong Kong Heritage Project, SEK-04-16, C.2-K-10/ (adjacent to /120).
5. S. Jasanoff and S. H. Kim, *Dreamscapes of Modernity: Sociotechnical Imaginaries and the Fabrication of Power* (Chicago: University of Chicago Press, 2015).

BIBLIOGRAPHY

ARCHIVES

Government Records Service, Public Records Office, Hong Kong
Hong Kong Heritage Project, Hong Kong
Hong Kong Public Libraries, Hong Kong Central Library (Carl T. Smith Collection), Hong Kong
Margaret Thatcher Foundation Archive (online; digital copies lodged at Thatcher Archive, Churchill College, Cambridge, United Kingdom)
National Archives, United Kingdom

PUBLICATIONS

Abbas, M. A. *Hong Kong: Culture and the Politics of Disappearance*. Hong Kong: Hong Kong University Press, 1997.
Aijmer, G. *Atomistic Society in Sha Tin: Immigrants in a Hong Kong Valley*. Göteborg: Acta Universitatis Gothoburgensis, 1986.
———. *Economic Man in Sha Tin: Vegetable Gardeners in a Hong Kong Valley*. London: Curzon Press, 1980.
Airlie, S. *Scottish Mandarin: The Life and Times of Sir Reginald Johnston*. Hong Kong: Hong Kong University Press, 2012.
———. *Thistle and Bamboo: The Life and Times of Sir James Stewart Lockhart*. Hong Kong: Oxford University Press, 1989.

BIBLIOGRAPHY

Akers-Jones, D. *Feeling the Stones: Reminiscences by David Akers-Jones.* Hong Kong: Hong Kong University Press, 2004.
Al, S., ed. *Mall City: Hong Kong's Dreamworlds of Consumption.* Hong Kong: Hong Kong University Press, 2016.
Allsop, A., ed. *St. George's Building: A Brief Portrait.* Hong Kong: Hong Kong Heritage Project, 2016.
Amatori, F., and G. Jones. *Business History Around the World.* Cambridge: Cambridge University Press, 2003.
Amsden, A. H. *Asia's Next Giant: South Korea and Late Industrialization.* New York: Oxford University Press, 1989.
Amsden, A. H., and W. W. Chu. *Beyond Late Development: Taiwan's Upgrading Policies.* Cambridge, MA: MIT Press, 2003.
Anderson, B. *Imagined Communities: Reflections on the Origin and Spread of Nationalism.* Revised edition. London: Verso, 2006.
Anderson, W. "Asia as Method in Science and Technology Studies." *East Asian Science, Technology and Society* 6, no. 4 (2012): 445–51.
Anslow, B. *Tin Hats & Rice: A Diary of Life as a Hong Kong Prisoner of War, 1941–1945.* Hong Kong: Blacksmith Books, 2018.
Armstrong, R., et al. *Castle Peak B Power Station Hong Kong Ex 81 (1).* (Ref A04046). 1981. http://www.Margaretthatcher.org/Document/138340.
Arnold, D. *Everyday Technology: Machines and the Making of India's Modernity.* Chicago: University of Chicago Press, 2013.
Arthur, W. B. *The Nature of Technology: What It Is and How It Evolves.* New York: Free Press, 2009.
Baark, E. *Lightning Wires: The Telegraph and China's Technological Modernization, 1860–1890.* Westport, CT: Greenwood Press, 1997.
Baker, B. E. *Shanghai: Electric and Lurid City: An Anthology.* Hong Kong: Oxford University Press, 1998.
Banham, T. *Not the Slightest Chance: The Defence of Hong Kong, 1941.* Hong Kong: Hong Kong University Press, 2003.
Bard, S. *Voices From the Past: Hong Kong, 1842–1918.* Hong Kong: Hong Kong University Press, 2002.
Barnard, T. P. *Nature Contained: Environmental Histories of Singapore.* Singapore: NUS Press, 2014.
Barrie, R. *Shares in Hong Kong: One Hundred Years of Stock Exchange Trading.* Hong Kong: The Stock Exchange of Hong Kong, 1991.
Bartlett, F. *The Peninsula: Portrait of a Grand Old Lady.* Hong Kong: Roundhouse Publications, 1997.
Beckert, J., and M. Zafirovski, eds. *International Encyclopedia of Economic Sociology.* London: Routledge, 2006.
Bellis, D. *Old Hong Kong Photos and the Tales They Tell.* Hong Kong: Gwulo, 2017.
Benson, S. *The Little World.* London: Macmillan, 1925.
Berger, S., and R. K. Lester, eds. *Made by Hong Kong.* Hong Kong: Oxford University Press, 1997.
Bevir, M. *Governance: A Very Short Introduction.* Oxford: Oxford University Press, 2012.

BIBLIOGRAPHY

Bew, J. *Clement Atlee: The Man Who Made Modern Britain.* Oxford: Oxford University Press, 2017.

Bickers, R., and Y. Yep, eds. *May Days in Hong Kong: Riot and Emergency in 1967.* Hong Kong: Hong Kong University Press, 2009.

Bickers, R. A. *Empire Made Me: An Englishman Adrift in Shanghai.* London: Allen Lane, 2003.

Bijker, W. E. *Of Bicycles, Bakelites, and Bulbs: Toward a Theory of Sociotechnical Change.* Cambridge, MA: MIT Press, 1995.

———. *The Social Construction of Technological Systems: New Directions in the Sociology and History of Technology.* Cambridge, MA: MIT Press, 2012.

Black, R. B. *Immigration and Social Planning in Hong Kong.* London: China Society, 1965.

Blackie, W. J. *Kadoorie Agricultural Aid Association, 1951–1971.* Hong Kong: Kadoorie Agricultural Aid Association, 1972.

Blyth, S., and I. Wotherspoon. *Hong Kong Remembers.* Hong Kong: Oxford University Press, 1996.

Bodanis, D. *Electric Universe: The Shocking True Story of Electricity.* New York: Crown Publishers, 2005.

Boon, G. C. *Technology and Entrepôt Colonialism in Singapore, 1819–1940.* Singapore: Institute of Southeast Asian Studies, 2013.

Bowers, B. *A History of Electric Light and Power.* Stevenage: Peregrinus, 1982.

———. *Lengthening the Day: A History of Lighting Technology.* Oxford: Oxford University Press, 1998.

Bowring, P. *Free Trade's First Missionary: Sir John Bowring in Europe and Asia.* Hong Kong: Hong Kong University Press, 2014.

Bradley, J. *The Imperial Cruise: A Secret History of Empire and War.* New York: Little, Brown, 2009.

Braga, J. M., ed. *Hong Kong Business Symposium: A Compilation of Authoritative Views on the Administration, Commerce and Resources of Britain's Far East Outpost.* Hong Kong: South China Morning Post, 1957.

Braga, S. "Making Impressions: The Adaptation of a Portuguese Family to Hong Kong, 1700–1950." PhD dissertation, Australian National University, 2012.

British Council. *English Books 1480–1940: An Exhibition of Rare Books Opened by His Excellency Sir Alexander Grantham, Governor of Hong Kong, in the Library of the British Council, Statue Square, Hong Kong, on the 14th October, 1949.* Hong Kong: British Council, 1949.

Cameron, N. *The Hongkong Land Company Ltd.: A Brief History.* Hong Kong: Privately published, 1979.

———. *An Illustrated History of Hong Kong.* Hong Kong: Oxford University Press, 1991.

———. *The Mandarin, Hong Kong.* Hong Kong: Mandarin International Hotels, 1980.

———. *The Milky Way: The History of Dairy Farm.* Hong Kong: Dairy Farm, 1986.

———. *Power: The Story of China Light.* Hong Kong: Oxford University Press, 1982.

Caro, R. A. *The Path to Power.* New York: Knopf, 1982.

Carroll, J. M. *A Concise History of Hong Kong.* Hong Kong: Hong Kong University Press, 2007.

BIBLIOGRAPHY

——. *Edge of Empires: Chinese Elites and British Colonials in Hong Kong*. Cambridge, MA: Harvard University Press, 2005.
Castells, M. *The Rise of the Network Society*. Chichester, UK: Wiley-Blackwell, 2010.
Cesarani, D. "An Alien Concept? The Continuity of Anti-Alienism in British Society Before 1940." In *The Internment of Aliens in Twentieth Century Britain*, edited by D. Cesarini and T. Kushner, 25–52. London: Frank Cass, 1993.
Chadwick, E. *Report on the Sanitary Condition of the Labouring Population of Gt. Britain*. Edinburgh: Edinburgh University Press, 1965.
Chandler, A. D., and B. Mazlish, eds. *Leviathans: Multinational Corporations and the New Global History*. Cambridge: Cambridge University Press, 2005.
Cheng, P. H., and P. M. Toong. *Kowloon Roads and Streets*. Hong Kong: Joint Publishing, 2003.
Cheong-leen, H. *Hongkong Tomorrow: A Collection of Speeches and Articles by Hilton Cheong-leen*. Hong Kong: Local Property & Printing, 1962.
Chernow, R. *Titan: The Life of John D Rockefeller, Sr.* New York: Vintage Books, 2004.
Cheung, G. K. *Hong Kong's Watershed: The 1967 Riots*. Hong Kong: Hong Kong University Press, 2009.
China Light & Power Co. (CLP). *Golden Memories of Tsing Yi Power Station*. Hong Kong: CLP, 1999.
——. *Large Thermal Power Station Site Search: Presentation to Tuen Mun District Board, Summary of the Site Search Study*. Hong Kong: CLP, 1990.
——. *Memorandum and New Articles of Association of The China Light & Power Co. (1918), Ltd*. Hong Kong: J. P. Braga, 1927.
——. *Power for Prosperity: An Overview of China Light & Power and its Role in Hong Kong's Development*. Hong Kong: CLP, 1991.
——. *Statistical Report, First October, 1958 to Thirtieth September, 1963*. Hong Kong: CLP, 1964.
——. *Submissions and Final Address of the China Light & Power Co., Ltd., to the Electricity Supply Companies Commission 1959*. Hong Kong: CLP, 1959.
China Mail. *Who's Who in the Far East, 1906–7*. Hong Kong: China Mail, 1906.
Ching, F. *The Li Dynasty: Hong Kong Aristocrats*. Hong Kong: Oxford University Press, 1999.
Clark, A. D., ed. *100 Years of Electricity*. Hong Kong: Hong Kong Electric Co., 1990.
Clifford, M. L. *The Greening of Asia: The Business Case for Solving Asia's Environmental Emergency*. New York: Columbia University Press, 2015.
——. *Troubled Tiger: Businessmen, Bureaucrats and Generals in South Korea*. Armonk, NY: M. E. Sharpe, 1994.
Clifford, M. L., and P. Engardio. *Meltdown: Asia's Spectacular Boom and Devastating Bust*. Parnassus, NJ: Prentice-Hall, 2000.
Coates, A. *China Races*. Hong Kong: Oxford University Press, 1983.
——. *The Commerce in Rubber: The First 250 Years*. Singapore: Oxford University Press, 1987.
——. *A Mountain of Light: The Story of the Hongkong Electric Company*. London: Heinemann, 1977.
——. *Myself a Mandarin: Memoirs of a Special Magistrate*. Hong Kong: Heinemann, 1975.
——. *Quick Tidings of Hong Kong*. Hong Kong: Oxford University Press, 1990.

———. *Whampoa: Ships on the Shore*. Hong Kong: South China Morning Post, 1980.
Cohen, P. A. *Discovering History in China: American Historical Writing on the Recent Chinese Past*. New York: Columbia University Press, 1984.
Coleman, L. *A Moral Technology: Electrification as Political Ritual in New Delhi*. Ithaca, NY: Cornell University Press, 2017.
Coll, S. *Private Empire: ExxonMobil and American Power*. New York: Penguin Books, 2012.
Collins, C. *Public Administration in Hong Kong*. London: Royal Institute of International Affairs, 1952.
Collis, M. *Wayfoong: The Hong Kong and Shanghai Banking Corporation: A Study of East Asia's Transformation, Political, Financial and Economic, During the Last Hundred Years*. London: Faber, 1965.
Coopersmith, J. *The Electrification of Russia, 1880–1926*. Ithaca, NY: Cornell University Press, 1992.
Cunich, P. *A History of the University of Hong Kong*. Hong Kong: Hong Kong University Press, 2012.
Darnton, R. *The Great Cat Massacre and Other Episodes in French Cultural History*. New York: Basic Books, 1984.
Davis, M. C. *Hongkong Tramways: 100 Years*. Croydon: DTS Publishing, 2004.
DeLanda, M. *Assemblage Theory*. Edinburgh: Edinburgh University Press, 2016.
———. *A New Philosophy of Society: Assemblage Theory and Social Complexity*. London: Continuum, 2006.
Des Voeux, G. W. *My Colonial Service in British Guiana, St. Lucia, Trinidad, Fiji, Australia, Newfoundland, and Hong Kong, with Interludes*. London: John Murray, 1903.
Dikötter, F. *The Age of Openness: China Before Mao*. Hong Kong: Hong Kong University Press, 2008.
———. *The Cultural Revolution: A People's History, 1962–1976*. New York: Bloomsbury Press, 2016.
———. *Exotic Commodities: Modern Objects and Everyday Life in China*. New York: Columbia University Press, 2006.
———. *Mao's Great Famine: The History of China's Most Devastating Catastrophe, 1958–1962*. New York: Bloomsbury Press, 2010.
———. *The Tragedy of Liberation: A History of the Chinese Revolution, 1945–1957*. New York: Bloomsbury Press, 2013.
Drage, C. *Two-Gun Cohen*. London: Jonathan Cape, 1954.
Edgerton, D. *The Shock of the Old: Technology and Global History Since 1900*. Oxford: Oxford University Press, 2011.
Elliot [Tu], E. *Colonial Hong Kong in the Eyes of Elsie Tu*. Hong Kong: Hong Kong University Press, 2003.
Emerson, G. C. *Hong Kong Internment, 1942 to 1945: Life in the Japanese Civilian Camp at Stanley*. Hong Kong: Hong Kong University Press, 2008.
Endacott, G. B. *A Biographical Sketch-Book of Early Hong Kong*. Singapore: Eastern Universities Press, 1962.
———. *An Eastern Entrepot: A Collection of Documents Illustrating the History of Hong Kong*. London: HMSO, 1964.
———. *Fragrant Harbour: A Short History of Hong Kong*. Hong Kong: Oxford University Press, 1962.

BIBLIOGRAPHY

———. *A History of Hong Kong*. Hong Kong: Oxford University Press, 1973.
———. *Hong Kong Eclipse*. Hong Kong: Oxford University Press, 1978.
———. *They Lived in Government House*. Hong Kong: Hong Kong University Press, 1981.
England, V. *Arnholds: China Trader*. Hong Kong: Arnhold & Co., 2017.
———. *The Croucher Foundation: The First 25 Years*. Hong Kong: The Croucher Foundation, 2004.
———. *Kindred Spirits: A History of the Hong Kong Club*. Hong Kong: Hong Kong Club, 2016.
———. *The Quest of Noel Croucher: Hong Kong's Quiet Philanthropist*. Hong Kong: Hong Kong University Press, 1998.
Ernst & Whinney. *Consultancy to Review the Government's Monitoring Arrangements of the Power Companies*. Hong Kong: Ernst & Whinney, 1984.
———. *Large Thermal Power Station Site Search: Presentation to District Boards: Summary of the Site Search Study*. Hong Kong: Ernst & Whinney, 1990.
Ewen, S. *What Is Urban History?* Cambridge: Polity Press, 2016.
Fan, Shu Ching. *The Population of Hong Kong*. Paris: Committee for International Coordination of National Research in Demography, 1974.
Far Eastern Economic Review. *Far Eastern Economic Review: Telling Asia's Story for Fifty Years*. Hong Kong: Far Eastern Economic Review, 1996.
Fellows, J. "Crafting Hong Kong's Image Overseas: 'Commercial Public Relations' in Hong Kong, 1962–1966." Unpublished paper, University of Hong Kong Spring History Symposium, 2016.
Ford, S. " 'Reel Sisters' and Other Diplomacy: Cathay Studios and Cold War Cultural Production." In *Hong Kong in the Cold War*, edited by P. Roberts and J. Carroll, 183–210. Hong Kong: Hong Kong University Press, 2016.
Forgan, S. "From Modern Babylon to White City: Science, Technology, and Urban Change in London, 1870–1914." In *Urban Modernity: Cultural Innovation in the Second Industrial Revolution*, edited by M. R. Levin et al., 75–132. Cambridge, MA: MIT Press, 2010.
Fortescue, D. *The Survivors: A Period Piece*. London: Anima Books, 2015.
Foucault, M. *Discipline and Punish: The Birth of the Prison*. London: Allen Lane, 1977.
Freese, B. *Coal: A Human History*. Cambridge, MA: Perseus, 2003.
French, P. *Bloody Saturday: Shanghai's Darkest Day*. Melbourne: Penguin Books, 2017.
———. *Through the Looking Glass: China's Foreign Journalists from Opium Wars to Mao*. Hong Kong: Hong Kong University Press, 2009.
Friedman, M. *Capitalism and Freedom*. Chicago: University of Chicago Press, 1962.
Friedman, M., and R. Friedman. *Free to Choose: A Personal Statement*. Harmondsworth, UK: Penguin Books, 1980.
Fruin, W. M. *Kikkoman: Company, Clan, and Community*. Cambridge, MA: Harvard University Press, 1983.
Fung, C. M. *Reluctant Heroes: Rickshaw Pullers in Hong Kong and Canton, 1874–1954*. Hong Kong: Hong Kong University Press, 2005.
Fung, K. *Upon the Plinth of a Barren Rock: 130 Years Engineering Development in Hong Kong*. Hong Kong: Chung Hwa, 2015.
Gerschenkron, A. *Economic Backwardness in Historical Perspective*. Cambridge, MA: Harvard University Press, 1962.
Gimson, F. "Internment in Hongkong." Unpublished manuscript.

BIBLIOGRAPHY

Girard, G. *HK: PM, Hong Kong Night Life 1974–1989*. Hong Kong: Asia One Publishing, 2017.

Gittins, J. *Eastern Windows—Western Skies*. Hong Kong: South China Morning Post, 1969.

Goh, C. B. *Technology and Entrepôt Colonialism in Singapore, 1819–1940*. Singapore: Institute of Southeast Asian Studies, 2014.

Goodstadt, L. F. *Profits, Politics and Panics: Hong Kong's Banks and the Making of a Miracle Economy, 1935–1985*. Hong Kong: Hong Kong University Press, 2007.

———. *Uneasy Partners: The Conflict Between Public Interest and Private Profit in Hong Kong*. Hong Kong: Hong Kong University Press, 2005.

Gordon, R. J. *The Rise and Fall of American Growth*. Princeton, NJ: Princeton University Press, 2016.

Government of the United Kingdom. *Index to CO 129 (1842–1951)*. London: Colonial Office, n.d.

———. *The War Dead of the British Commonwealth and Empire: The Register of the Names of Those Who Fell in the 1939–1945 War and Are Buried in Cemeteries in Hong Kong*. London: Imperial War Graves Commission, 1956.

Graham, L. R. *The Ghost of the Executed Engineer: Technology and the Fall of the Soviet Union*. Cambridge, MA: Harvard University Press, 1993.

Grantham, A. W. "Hong Kong." *Journal of the Royal Central Asian Society* 46, no. 2 (1959): 119–29.

———. *The University of Hong Kong: An Address to the Court of the University, at Its Meeting on Wednesday, 7th April, 1948*. Hong Kong: South China Morning Post, 1948.

———. *Via Ports, from Hong Kong to Hong Kong*. Hong Kong: Hong Kong University Press, 1965.

———. *Via Ports: From Hong Kong to Hong Kong*. Revised edition. Hong Kong: Hong Kong University Press, 2012.

Hamilton, G. G. *Asian Business Networks*. Berlin: Walter de Gruyter, 1996.

———. *Cosmopolitan Capitalists: Hong Kong and the Chinese Diaspora at the End of the 20th Century*. Seattle: University of Washington Press, 1999.

Hannah, L. *British Electricity Before Nationalisation: A Study of the Development of the Electricity Supply Industry in Britain to 1948*. London: Macmillan, 1979.

Harrington, A. M., L. Barbara, and H. P. Müller. *Encyclopedia of Social Theory*. London: Routledge, 2006.

Hausman, W. J., P. Hertner, and M. Wilkins. *Global Electrification: Multinational Enterprise and International Finance in the History of Light and Power, 1878–2007*. Cambridge: Cambridge University Press, 2008.

Headrick, D. R. *The Tools of Empire: Technology and European Imperialism in the Nineteenth Century*. New York: Oxford University Press, 1981.

Helm, Dieter. *Energy, the State, and the Market: British Energy Policy Since 1979*. Oxford: Oxford University Press, 2003.

Henry, T. A. *Assimilating Seoul: Japanese Rule and the Politics of Public Space in Colonial Korea, 1910–1945*. Berkeley: University of California Press, 2013.

Hibbard, P. *Beyond Hospitality: The History of the Hongkong and Shanghai Hotels, Limited*. Hong Kong: Hongkong and Shanghai Hotels, 2010.

Ho, K. W. "A Comparative Study of the Corporate Strategies of the Two Electricity Companies in Hong Kong." MBA thesis, University of Hong Kong, 1992.

Ho, V. *Understanding Canton: Rethinking Popular Culture in the Republican Period.* Oxford: Oxford University Press, 2005.
Hoe, S. *The Private Life of Old Hong Kong: Western Women in the British Colony, 1841–1941.* Hong Kong: Oxford University Press, 1991.
Holdsworth, M., and C. Munn, eds. *Dictionary of Hong Kong Biography.* Hong Kong: Hong Kong University Press, 2012.
Hong Kong and China Gas Co. *Lighting the Past, Brightening the Future.* Hong Kong: Hong Kong and China Gas Co., 2012.
Hong Kong Government. *Administration Report,* 1879–1941.
———. *Annual Report,* 1879–1986.
———. *Commission of Inquiry Into Matters Relating to the Electric Companies.* Hong Kong: Government Printer, 1959.
———. Legislative Council, Panel on Economic Development. Public Consultation on the Future Development of the Electricity Market. 2015. http://www.legco.gov.hk/yr14-15/english/panels/edev/papers/edevcb4-727-1-e.pdf.
———. *A Problem of People.* Hong Kong: Government Printer, 1957.
———. *Report on the Riots in Kowloon and Tsuen Wan, October 10th to 12th, 1956, Together with Covering Despatch Dated the 23rd December, 1956, from the Governor of Hong Kong to the Secretary of State for the Colonies.* Hong Kong: Government Printer, 1956.
———. *Report of the Working Party of Electrification of Squatter Areas.* Hong Kong: Government Printer, 1976.
Hong Kong Heritage Project. *One Hundred Years of Kadoorie Business Records at a Glance.* Hong Kong: Hong Kong Heritage Project, 2011.
Hou, C. *Foreign Investment and Economic Development in China: 1840–1937.* New York: Routledge, 2000.
Hughes, R. *Borrowed Place, Borrowed Time: Hong Kong and Its Many Faces.* London: Andre Deutsch, 1976.
Hughes, T. *American Genesis: A Century of Invention and Technological Enthusiasm, 1870–1970.* Chicago: University of Chicago Press, 2004.
———. *The Development of Western Technology Since 1500.* New York: Macmillan, 1964.
———. *Human-Built World: How to Think About Technology and Culture.* Chicago: University of Chicago Press, 2004.
———. *Networks of Power: Electrification in Western Society, 1880–1930.* Baltimore, MD: Johns Hopkins University Press, 1983.
———. *Rescuing Prometheus.* New York: Pantheon Books, 1998.
Hunt, T. *Ten Cities That Made an Empire.* London: Allen Lane, 2014.
Hutcheon, R. *The Blue Flame: 125 Years of Towngas in Hong Kong.* Hong Kong: Hong Kong & China Gas, 1987.
———. *Burst of Crackers: The Li & Fung Story.* Hong Kong: Li & Fung, 1992.
———. *First Sea Lord: The Life and Work of Sir Y. K. Pao.* Hong Kong: Chinese University Press, 1990.
———. *High-Rise Society: The First 50 Years of the Hong Kong Housing Society.* Hong Kong: Hong Kong Housing Society, 1998.
———. *The Merchants of Shameen: The Story of Deacon & Co.* Hong Kong: Deacon, 1990.
———. *SCMP: The First Eighty Years.* Hong Kong: South China Morning Post, 1983.
———. *Wharf: The First Hundred Years.* Hong Kong: Wharf Holdings, 1986.

BIBLIOGRAPHY

Hyam, R., and W. R. Louis. *The Conservative Government and the End of Empire, 1957–1964, Part I: High Policy, Political and Constitutional Change.* London: The Stationery Office, 2000.

———. *The Conservative Government and the End of Empire, 1957–1964, Part II: Economics, International Relations, and the Commonwealth.* London: The Stationery Office, 2000.

Imber, E. E. "A Late Imperial Elite Jewish Politics: Baghdadi Jews in British India and the Political Horizons of Empire and Nation." *Jewish Social Studies* 23, no. 2 (2018): 48–85.

Jackson, S. *The Sassoons.* New York: Dutton, 1968.

Jacobs, M. *Panic at the Pump: The Energy Crisis and the Transformation of American Politics in the 1970s.* New York: Hill and Wang, 2016.

James, W. "Hong Kong—Boom Town." *Collier's Weekly*, January 10, 1948, 52–57.

Jao, Y. C. *Banking and Currency in Hong Kong: A Study of Postwar Financial Development.* London: Macmillan, 1974.

———. "Financing Hong Kong's Early Postwar Industrialization: The Role of the Hongkong and Shanghai Banking Corporation." In *Eastern Banking: Essays in the History of the Hongkong and Shanghai Banking Corporation*, edited by F. H. King, 545–74. London: Athlone Press, 1983.

Jasanoff, S. "Future Imperfect: Science, Technology, and the Imaginations of Modernity." In *Dreamscapes of Modernity: Sociotechnical Imaginaries and the Fabrication of Power*, edited by S. Jasanoff and S. H. Kim, 1–33. Chicago: University of Chicago Press, 2015.

———. "Imagined and Invented Worlds." In Jasanoff and Kim, *Dreamscapes of Modernity*, 321–41.

———. *Science and Public Reason.* Milton Park: Routledge, 2012.

———, ed. *States of Knowledge: The Co-Production of Science and Social Order.* London: Routledge, 2004.

Jones, G. *Merchants to Multinationals: British Trading Companies in the Nineteenth and Twentieth Centuries.* Oxford: Oxford University Press, 2000.

Jones, M. "Tuberculosis, Housing and the Colonial State: Hong Kong, 1900–1950." *Modern Asian Studies* 37, no. 3 (2003): 653–82.

Jonnes, J. *Empires of Light: Edison, Tesla, Westinghouse, and the Race to Electrify the World.* New York: Random House, 2003.

Kadoorie, L. "Chairman's Speech for Inaugural Ceremony of New Power Station." Hong Kong: CLP, 1940.

———. "China Light & Power Co. Ltd.: 1901–1918." Hong Kong: CLP, 1977.

———. "Electricity Success Story." *Financial Times*, July 19, 1965.

———. "General Policy (A Brief Survey)." Unpublished manuscript, 1938.

———. "Hong Kong Jewish Community." Unpublished manuscript, 1946.

———. "1997 and All That." Unpublished manuscript, 1973.

———. "1997 and All That." Unpublished manuscript, 1976.

———. "1997 and All That—Further Thoughts." Unpublished manuscript, 1975.

Kale, S. S. *Electrifying India: Regional Political Economies of Development.* Stanford, CA: Stanford University Press, 2014.

Kaufman, J. *The Last Kings of Shanghai: The Rival Jewish Dynasties That Helped Create Modern China.* New York: Viking Press, 2020.

Keswick, J. "The Romance of Trade in Hong Kong." In Braga, *Hong Kong Business Symposium*, 33–38.
Keswick, M. *The Thistle and the Jade: A Celebration of 150 Years of Jardine, Matheson & Co.* London: Octopus Books, 1982.
Kikkawa, T. "The History of Japan's Electric Power Industry Before World War II." *Hitotsubashi Journal of Commerce and Management* 46 (2012): 1–16.
Killen, A. *Berlin Electropolis: Shock, Nerves, and German Modernity*. Berkeley: University of California Press, 2006.
King, F. H., ed. *Eastern Banking: Essays in the History of the Hongkong and Shanghai Banking Corporation*. London: Athlone Press, 1983.
King, F. H. H. *The Hongkong Bank Between the Wars and the Bank Interned, 1919–1945*. Cambridge: Cambridge University Press, 1998.
———. *The Hongkong Bank in Late Imperial China, 1864–1902*. Cambridge: Cambridge University Press, 1987.
———. *The Hongkong Bank in the Period of Development and Nationalism, 1941–1984*. Cambridge: Cambridge University Press, 1991.
———. *The Hongkong Bank in the Period of Imperialism and War, 1895–1918*. Cambridge: Cambridge University Press, 1988.
King, F. H. H., and P. Clark. *A Research Guide to China-Coast Newspapers, 1822–1911*. Cambridge, MA: East Asian Research Center, Harvard University, 1965.
Ko, T. K., and J. Wordie. *Ruins of War: A Guide to Hong Kong's Battlefields and Wartime Sites*. Hong Kong: Joint Publishing, 1996.
Koeppel, G. T. *Water for Gotham: A History*. Princeton, NJ: Princeton University Press, 2000.
Kong, V. W. Y. "'Clearing the Decks': The Evacuation of British Women and Children from Hong Kong to Australia in 1940." MPhil thesis, University of Hong Kong, 2015.
Kong, Y. C. "Jewish Merchants' Community in Shanghai: A Study of the Kadoorie Enterprise, 1890–1950." PhD dissertation, Hong Kong Baptist University, 2017.
Kuok, R., and A. Tanzer. *A Memoir*. Singapore: Landmark Books, 2017.
Kwok, S. T. *A Century of Light*. Hong Kong: CLP, 2001.
Kwong, C. M., and Y. L. Tsoi. *Eastern Fortress: A Military History of Hong Kong, 1840–1970*. Hong Kong: Hong Kong University Press, 2014.
Lam, P. L. *The Scheme of Control on Electricity Companies*. Hong Kong: Chinese University Press, 1996.
Lam, P. L., and S. Chan. *Competition in Hong Kong's Gas Industry*. Hong Kong: Chinese University Press, 2000.
Lamb, C. K. "Supply Companies Ahead of Hong Kong Industry." *Asian Industry*, April 1966, 11–12.
Lambert, J. *The Power Brokers: The Struggle to Shape and Control the Electric Power Industry*. Cambridge, MA: MIT Press, 2015.
Lambot, I., and G. Chambers. *One Queen's Road Central*. Hong Kong: Hongkong Bank, 1986.
Landes, D. *The Unbound Prometheus: Technological Change and Industrial Development in Western Europe from 1750 to the Present*. Cambridge: Cambridge University Press, 2003.
Latour, B. *Aramis, or the Love of Technology*. Cambridge, MA: Harvard University Press, 1996.

BIBLIOGRAPHY

———. *The Making of Law: An Ethnography of the Conseil d'Etat*. Cambridge: Polity, 2010.
———. *Reassembling the Social: An Introduction to Actor-Network-Theory*. Oxford: Oxford University Press, 2005.
———. *Science in Action: How to Follow Scientists and Engineers Through Society*. Milton Keynes: Open University Press, 1987.
———. *We Have Never Been Modern*. Cambridge, MA: Harvard University Press, 1993.
Lee, J. S., ed. *The Emergence of the South China Growth Triangle*. Taipei: Chung-Hua Institution for Economic Research, 1996.
Lenin, V. I. "Our Foreign and Domestic Position and Party Tasks: Speech Delivered to the Moscow Gubernia Conference of the R. C. P. (B.)." 1920. https://www.marxists.org/archive/lenin/works/1920/nov/21.htm.
Lent, J. A. *The Asian Newspapers' Reluctant Revolution*. Ames: Iowa State University Press, 1971.
Lester, A. "Empire and the Place of Panic." In *Empires of Panic: Epidemics and Colonial Anxieties*, edited by R. Peckham, 23–34. Hong Kong: Hong Kong University Press, 2015.
Leung, K. W. "The Strategic Importance of Information Systems in the Electricity Supply Industry in Hong Kong." MBA thesis, University of Hong Kong Business School, 1995.
Leventhal, D., and L. Kadoorie. *Sino-Judaic Studies: Whence and Whither: An Essay and Bibliography and the Kadoorie Memoir*. Hong Kong: Hong Kong Jewish Chronicle, 1985.
Levin, M. R. *Cultures of Control*. Amsterdam: Abingdon, 2000.
Levin, M. R., S. Forgan, and M. Hessler, et al. *Urban Modernity: Cultural Innovation in the Second Industrial Revolution*. Cambridge, MA: MIT Press, 2010.
Li, K. W. *Economic Freedom: Lessons of Hong Kong*. Singapore: World Scientific Publishing, 2012.
Li, S. F. *Hong Kong Surgeon*. London: Victor Gollancz, 1964.
Liberman, P. *Does Conquest Pay? The Exploitation of Occupied Industrial Societies*. Princeton, NJ: Princeton University Press, 1996.
Lieberman, J. L. *Power Lines: Electricity in American Life and Letters, 1882–1952*. Cambridge, MA: MIT Press, 2017.
Lim, P. *Forgotten Souls: A Social History of the Hong Kong Cemetery*. Hong Kong: Hong Kong University Press, 2011.
Lloyd, E., and A. Licata. *One New York City: One Water*. New York: New York City Department of Environmental Protection, n.d.
Loh, C. *Underground Front: The Chinese Communist Party in Hong Kong*. Hong Kong: Hong Kong University Press, 2010.
Louis, W. R. "Hong Kong: The Critical Phase, 1945–1949." *American Historical Review* 102, no. 4 (1997): 1057–84.
MacKenzie, D. A., and J. Wajcman. *The Social Shaping of Technology: How the Refrigerator Got Its Hum*. Milton Keynes: Open University Press, 1985.
MacPherson, K. L. *A Wilderness of Marshes: The Origins of Public Health in Shanghai, 1843–1893*. Hong Kong: Oxford University Press, 1987.
Marx, L. *The Machine in the Garden: Technology and the Pastoral Ideal in America*. New York: Oxford University Press, 1964.

Mathews, G. *Ghetto at the Center of the World*. Hong Kong: Hong Kong University Press, 2011.

Maunder, W. F. *Hong Kong Housing Survey, 1957*. Hong Kong: Special Committee on Housing, Hong Kong Government, 1957.

McDonald, F. *Insull*. Chicago: University of Chicago Press, 1962.

McLaren, R. *Britain's Record in Hong Kong*. London: Royal Institute of International Affairs, 1997.

Meyer, M. J. "Baghdadi Jewish Merchants in Shanghai and the Opium Trade." *Jewish Culture and History* 2, no. 1 (1999): 58–71.

———. *From the Rivers of Babylon to the Whangpoo: A Century of Sephardi Jewish Life in Shanghai*. Lanham, MD: University Press of America, 2003.

———. *Shanghai's Baghdadi Jews: A Collection of Biographical Reflections*. Hong Kong: Blacksmith Books, 2015.

Miller, I. J., J. A. Thomas, and B. L. Walker, eds. *Japan at Nature's Edge: The Environmental Context of a Global Power*. Honolulu: University of Hawai'i Press, 2013.

Min, A. *Red Azalea*. New York: Anchor Books, 2006.

Minami, R. *Power Revolution in the Industrialization of Japan, 1885–1940*. Tokyo: Kinokuniya, 1987.

Minnick, S. *The Repulse Bay: A Life of Elegance and Charm*. Hong Kong: Repulse Bay Co., 2012.

Mizuoka, F. "Contriving '*Laissez-faire*': Conceptualising the British Colonial Rule of Hong Kong." *City, Culture and Society* 5, no. 1 (2014): 23–32.

Mokyr, J. *The Lever of Riches: Technological Creativity and Economic Progress*. New York: Oxford University Press, 1990.

Monks, S. *Toy Town: How a Hong Kong Industry Played a Global Game*. Hong Kong: Toys Manufacturers' Association of Hong Kong, 2010.

Monnery, N. *Architect of Prosperity: Sir John Cowperthwaite and the Making of Hong Kong*. London: London Publishing Partnership, 2017.

Moore, A. S. *Constructing East Asia: Technology, Ideology, and Empire in Japan's Wartime Era, 1931–1945*. Stanford, CA: Stanford University Press, 2013.

Moore, C. *Margaret Thatcher: The Authorized Biography*. 2 volumes. London: Allen Lane, 2013–2015.

Morris, J. *Hong Kong: Epilogue to an Empire*. New York: Vintage Books, 1997.

Morus, I. R. *Frankenstein's Children: Electricity, Exhibition, and Experiment in Early-Nineteenth-Century London*. Princeton, NJ: Princeton University Press, 1998.

———. *Michael Faraday and the Electrical Century*. Cambridge: Icon, 2004.

Mosk, C. *Japanese Industrial History: Technology, Urbanization, and Economic Growth*. Armonk, NY: M. E. Sharpe, 2001.

Munn, C. *Anglo-China: Chinese People and British Rule in Hong Kong, 1841–1880*. Richmond, UK: Curzon, 2001.

Nakano, Y. *Where There Are Asians, There Are Rice Cookers: How "National" Went Global via Hong Kong*. Hong Kong: Hong Kong University Press, 2009.

Newman, G. *A Survey of Design in Britain, 1915–1939*. Milton Keynes: Open University Press, 1975.

Ngo, T. W. *Hong Kong's History: State and Society Under Colonial Rule*. London: Routledge, 1999.

BIBLIOGRAPHY

Nye, D. E. *Electrifying America: Social Meanings of a New Technology, 1880–1940.* Cambridge, MA: MIT Press, 1990.
——. *How Does One Do the History of Technology?* Detroit, MI: The Society for the History of Technology, 2014.
——. *Technology Matters: Questions to Live With.* Cambridge, MA: MIT Press, 2006.
——. *When the Lights Went Out: A History of Blackouts in America.* Cambridge, MA: MIT Press, 2010.
Olson, M. *The Rise and Decline of Nations: Economic Growth, Stagflation, and Social Rigidities.* New Haven, CT: Yale University Press, 1982.
Onley, J. *The Arabian Frontier of the British Raj: Merchants, Rulers, and the British in the Nineteenth-Century Gulf.* New York: Oxford University Press, 2007.
Osterhammel, J., and N. P. Petersson. *Globalization: A Short History.* Princeton, NJ: Princeton University Press, 2005.
Ostrom, E. *Governing the Commons: The Evolution of Institutions for Collective Action.* Cambridge: Cambridge University Press, 1990.
Oxford, E. *At Least We Lived: The Unlikely Adventures of an English Couple in World War II China.* Charleston, SC: Branksome Books, 2013.
Pan, Lynn, and Trea Wiltshire. *Saturday's Child: Hong Kong in the Sixties.* Hong Kong: Form Asia, 1993.
Pang, W. N. "From Squatter Settlements to Industrial Estates: The Impact of the Resettlement Program on Squatter Factories in Hong Kong—A Study of the Locational Problems of Urban Small Industries." PhD dissertation, University of California, Los Angeles, 1973.
Pao, A. S. *Y. K. Pao, My Father.* Hong Kong: Hong Kong University Press, 2013.
Patrick, H., ed. *Japanese Industrialization and Its Social Consequences.* Berkeley: University of California Press, 1976.
Peckham, R. "Hygienic Nature: Afforestation and the Greening of Colonial Hong Kong." *Modern Asian Studies* 49, no. 4 (2015): 1177–1209.
Perelman, M. *The Invention of Capitalism: Classical Political Economy and the Secret History of Primitive Accumulation.* Durham, NC: Duke University Press, 2000.
Phipps, C. L. *Empires on the Waterfront: Japan's Ports and Power, 1858–1899.* Cambridge, MA: Harvard University Asia Center, 2015.
Plachno, L., ed. *The Memoirs of Samuel Insull.* Polo, IL: Transportation Trails, 1992.
Poulter, J. D. *An Early History of Electricity Supply: The Story of the Electric Light in Victorian Leeds.* London: P. Peregrinus, 1986.
Pryor, E. G. *Housing in Hong Kong.* Hong Kong: Oxford University Press, 1983.
Rabinbach, A. *The Human Motor: Energy, Fatigue, and the Origins of Modernity.* New York: Basic Books, 1990.
Rabushka, A. *Hong Kong: A Study in Economic Freedom.* Chicago: University of Chicago, 1979.
Rafferty, K. *City on the Rocks: Hong Kong's Uncertain Future.* London: Viking Press, 1989.
Read, D. *The Power of News: The History of Reuters.* Oxford: Oxford University Press, 1999.
Roach, C. R. *Simply Electrifying: The Technology That Transformed the World, from Benjamin Franklin to Elon Musk.* Dallas, TX: Ben Bella Books, 2017.

Roberts, P., and J. Carroll, eds. *Hong Kong in the Cold War*. Hong Kong: Hong Kong University Press, 2016.

Roland, J. G. *The Jewish Communities of India: Identity in a Colonial Era*. New Brunswick, NJ: Transaction Publishers, 1998.

Rosenberg, N. *Inside the Black Box: Technology and Economics*. Cambridge: Cambridge University Press, 1982.

Ross, K. *Fast Cars, Clean Bodies: Decolonization and the Reordering of French Culture*. Cambridge, MA: MIT Press, 1996.

Sayer, G. R. *Hong Kong, 1841–1862: Birth, Adolescence, and Coming of Age*. Hong Kong: Hong Kong University Press, 1980.

———. *Hong Kong, 1862–1919: Years of Discretion*. Hong Kong: Hong Kong University Press, 1975.

Sayer, R. *The Man Who Turned the Lights On: Gordon Wu*. Hong Kong: Chameleon Press, 2006.

Schaffer, S., and S. Shapin. *Leviathan and the Air-Pump: Hobbes, Boyle, and the Experimental Life*. Princeton, NJ: Princeton University Press, 1985.

Schell, O., and J. Delury. *Wealth and Power: China's Long March to the Twenty-First Century*. New York: Random House, 2013.

Schenk, C. R. *Hong Kong as an International Financial Centre: Emergence and Development 1945–1965*. London: Routledge, 2001.

———. *Hong Kong Banks and Industry 1950–1970*. Manchester: University of Manchester Press, 2002.

Schivelbusch, W. *Disenchanted Night: The Industrialization of Light in the Nineteenth Century*. Berkeley: University of California Press, 1995.

Scott, J. C. *Seeing Like a State: How Certain Schemes to Improve the Human Condition Have Failed*. New Haven, CT: Yale University Press, 1998.

Scott, R. L., Jr. *God Is My Co-Pilot*. New York: Charles Scribner's Sons, 1944.

Sellar, W. C., R. J. Yeatman, and J. Reynolds. *1066 and All That: A Memorable History of England, Comprising All the Parts You Can Remember, Including 103 Good Things, 5 Bad Kings and 2 Genuine Dates*. London: Methuen, 1951.

Selwyn-Clarke, S. *Footprints: The Memoirs of Sir Selwyn Selwyn-Clarke*. Hong Kong: Sino-American Publishing, 1975.

Shan, W. *Out of the Gobi: My Story of China and America*. Hoboken, NJ: Wiley, 2019.

Sharpe, W. C. *New York Nocturne: The City After Dark in Literature, Painting, and Photography*. Princeton, NJ: Princeton University Press, 2008.

Sheller, M. *Aluminum Dreams: The Making of Light Modernity*. Cambridge, MA: MIT Press, 2014.

Shewan, A. *The Great Days of Sail: Some Reminiscences of a Tea-Clipper Captain*. Boston: Houghton Mifflin, 1927.

Sinn, E. *Growing with Hong Kong: The Bank of East Asia 1919–1994*. Hong Kong: Bank of East Asia, 1994.

———. "Lesson in Openness: Creating a Space of Flow in Hong Kong." In *Hong Kong Mobile: Making a Global Population*, edited by H. F. Siu and A. S. Ku, 13–44. Hong Kong: Hong Kong University Press, 2008.

———. *Pacific Crossing: California Gold, Chinese Migration, and the Making of Hong Kong*. Hong Kong: Hong Kong University Press, 2014.

BIBLIOGRAPHY

Smart, A. *The Shek Kip Mei Myth: Squatters, Fires and Colonial Rule in Hong Kong, 1950–1963*. Hong Kong: University of Hong Kong Press, 2006.
Smil, V. *China's Environmental Crisis: An Inquiry into the Limits of National Development*. Armonk, NY: M. E. Sharpe, 1993.
———. *Power Density: A Key to Understanding Energy Sources and Uses*. Cambridge, MA: MIT Press, 2015.
Smith, M. R., and L. Marx, eds. *Does Technology Drive History? The Dilemma of Technological Determinism*. Cambridge, MA: MIT Press, 1994.
Snow, P. *The Fall of Hong Kong*. New Haven, CT: Yale University Press, 2003.
Soll, D. *Empire of Water: An Environmental and Political History of the New York City Water Supply*. Ithaca, NY: Cornell University Press, 2013.
Stein, S. A. "Protected Persons? The Baghdadi Jewish Diaspora, the British State, and the Persistence of Empire." *American Historical Review* 116, no. 1 (2011): 80–108.
Stock Exchange of Hong Kong. *A Decade of Challenge and Development*. Hong Kong: Stock Exchange of Hong Kong, 1996.
———. *Shares in Hong Kong*. Hong Kong: Stock Exchange of Hong Kong, 1991.
Strickland, J. E. *Southern District Officer Reports: Islands and Villages in Rural Hong Kong, 1910–60*. Hong Kong: Hong Kong University Press, 2010.
Stuart, A. *Of Cargoes, Colonies, and Kings: Diplomatic and Administrative Service from Africa to the Pacific*. London: Radcliffe Press, 2001.
Szczepanik, E. F. *The Economic Growth of Hong Kong*. London: Oxford University Press, 1958.
———. *The Embargo Effect on China's Trade with Hong Kong*. Hong Kong: Hong Kong University Press, 1958.
Tagliacozzo, E. "The Lit Archipelago: Coast Lighting and the Imperial Optic in Insular Southeast Asia, 1860–1910." *Technology and Culture* 46, no. 2 (2005): 306–28.
Tam Wai-chu, M., ed. *The Basic Law and Hong Kong: The 15th Anniversary of Reunification with the Motherland*. Hong Kong: Working Group on Overseas Community of the Basic Law Promotion Steering Committee, 2012.
Taus, E. "Electrical Appliances." In Braga, *Hong Kong Business Symposium*, 426.
Terry, E. *How Asia Got Rich: Japan, China and the Asian Miracle*. Armonk, NY: M. E. Sharpe, 2002.
Thompson, J. B. *Ideology and Modern Culture: Critical Social Theory in the Era of Mass Communication*. Cambridge: Polity Press, 1990.
Tobey, R. C. *Technology as Freedom: The New Deal and the Electrical Modernization of the American Home*. Berkeley: University of California Press, 1996.
Tsai, J. F. *Hong Kong in Chinese History: Community and Social Unrest in the British Colony, 1842–1913*. New York: Columbia University Press, 1993.
Tsang, S. *Governing Hong Kong: Administrative Officers from the Nineteenth Century to the Handover to China, 1862–1997*. Hong Kong: Hong Kong University Press, 2007.
———. *A Modern History of Hong Kong*. Hong Kong: Hong Kong University Press, 2004.
Tsin, M. T. *Nation, Governance, and Modernity in China: Canton, 1900–1927*. Stanford, CA: Stanford University Press, 1999.
Tsing, A. L. *Friction: An Ethnography of Global Connection*. Princeton, NJ: Princeton University Press, 2005.
Tsui, Y. C. *The Hong Kong Independent Battalion: Local Heroes, the Hong Kong Resistance During the Japanese Occupation, 1941–45*. Hong Kong: Man Lai Ting, 2018.

Tu, E. *Colonial Hong Kong in the Eyes of Elsie Tu*. Hong Kong: Hong Kong University Press, 2003.
Ure, C. J. "Elevators." In Braga, *Hong Kong Business Symposium*, 463.
Victor, D. G., and T. C. Heller, eds. *The Political Economy of Power Sector Reform: The Experiences of Five Major Developing Countries*. Cambridge: Cambridge University Press, 2007.
Vines, S. *The Story of St John's Cathedral*. Hong Kong: FormAsia, 2001.
Vitasoy. *Vitasoy: Seventy Years of Popularity, A Story Dedicated to Humanity*. Hong Kong: Vitasoy International Holdings, 2010.
Vogel, E. F. *Deng Xiaoping and the Transformation of China*. Cambridge, MA: Harvard University Press, 2011.
———. *One Step Ahead in China: Guangdong Under Reform*. Cambridge, MA: Harvard University Press, 1989.
Wakefield, F. J. "Hong Kong Coaling Agency." In Braga, *Hong Kong Business Symposium*, 407.
Walker, A., and S. Rowlinson. *The Building of Hong Kong: Constructing Hong Kong Through the Ages*. Hong Kong: Hong Kong University Press, 1990.
Wasik, J. F. *The Merchant of Power: Sam Insull, Thomas Edison, and the Creation of the Modern Metropolis*. New York: Palgrave Macmillan, 2006.
Waters, D. "Hong Kong Hongs with Long Histories and British Connections." *Journal of the Hong Kong Branch of the Royal Asiatic Society* 30 (1990): 219–56.
White, B. *Turbans and Traders: Hong Kong's Indian Communities*. Hong Kong: Oxford University Press, 1994.
Whitfield, A. J. *Hong Kong, Empire and the Anglo-American Alliance at War, 1941–45*. Hong Kong: Hong Kong University Press, 2001.
Wiens, H. J. "Riverine and Coastal Junks in China's Commerce." *Economic Geography* 31, no. 3 (1955): 248–64.
Williams, R. *Notes on the Underground: An Essay on Technology, Society, and the Imagination*. Cambridge, MA: MIT Press, 1990.
———. *Retooling: A Historian Confronts Technological Change*. Cambridge, MA: MIT Press, 2002.
———. *The Triumph of Human Empire: Verne, Morris, and Stevenson at the End of the World*. Chicago: University of Chicago Press, 2013.
Winner, L. "Do Artifacts Have Politics." *Daedalus* 109, no. 1 (1980): 121–36.
Winseck, D., and R. Pike. *Communication and Empire: Media, Markets, and Globalization, 1860–1930*. Durham, NC: Duke University Press, 2007.
Winther, T. *The Impact of Electricity: Development, Desires and Dilemmas*. New York: Berghahn Books, 2008.
Witkovsky, M. S., C. S. Eliel, and K. P. B. Vail, eds. *Moholy-Nagy: Future Present*. Chicago: The Art Institute of Chicago, 2016.
Wittfogel, K. A. *Oriental Despotism: A Comparative Study of Total Power*. New Haven, CT: Yale University Press, 1957.
Wittner, D. G. *Technology and the Culture of Progress in Meiji Japan*. London: Routledge, 2008.
Wong, J., and C. K. Wong. *China's Power Sector*. Singapore: World Scientific Publishing, 1999.

BIBLIOGRAPHY

Wong, K. F. *A Fishery Tale: Wholesale Fish Marketing in Hong Kong.* Hong Kong: Asia Case Research Centre, The University of Hong Kong, 2005.

Wong, S. L. "Chinese Entrepreneurs and Business Trust." In *Business Networks and Economic Development in East and Southeast Asia*, edited by G. G. Hamilton, 13–26. Hong Kong: University of Hong Kong, 1996.

——. *Emigrant Entrepreneurs: Shanghai Industrialists in Hong Kong.* Hong Kong: Oxford University Press, 1988.

Wong, W. S. B. "Air Conditioning." In Braga, *Hong Kong Business Symposium*, 388–89.

Wood, C. F. "China Light & Power Co., Ltd." In Braga, *Hong Kong Business Symposium*, 312–15.

World Bank. *The East Asian Miracle: Economic Growth and Public Policy.* Oxford: Oxford University Press, 1993.

Woronoff, J. *Hong Kong: Capitalist Paradise.* Hong Kong: Heinemann, 1980.

Wright, A., and H. A. Cartwright, eds. *Twentieth Century Impressions of Hongkong, Shanghai, and Other Treaty Ports of China: Their History, People, Commerce, Industries, and Resources.* London: Lloyd's Greater Britain Publishing, 1908.

Wu, S. X. *Empires of Coal: Fueling China's Entry Into the Modern World Order, 1860–1920.* Stanford, CA: Stanford University Press, 2015.

Wue, R. *Picturing Hong Kong: Photography, 1855–1910.* New York: Asia Society, 1997.

Xu, Y. C. *Sinews of Power: The Politics of the State Grid Corporation of China.* New York: Oxford University Press, 2017.

Yang, D. *Technology of Empire: Telecommunications and Japanese Expansion in Asia, 1883–1945.* Cambridge, MA: Harvard University Asia Center, 2010.

Yang, J. *Tombstone: The Great Chinese Famine, 1958–1962.* New York: Farrar, Straus and Giroux, 2013.

Yergin, D. *The Prize: The Epic Quest for Oil, Money, and Power.* New York: Simon & Schuster, 1991.

——. *The Quest: Energy, Security and the Remaking of the Modern World.* London: Allen Lane, 2011.

Young, L. *Beyond the Metropolis: Second Cities and Modern Life in Interwar Japan.* Berkeley: University of California Press, 2013.

Young, M. *Events in Hong Kong on 25th December, 1941.* Hong Kong: Government Gazette, 1948.

Zhang, L. *China Remembers.* Hong Kong: Oxford University Press, 1999.

INDEX

AC (alternating current), 123
Administrative Reports, 39–40, 51, 177
Advisory Committee on Education in the Colonies, 30
Aikawa Haruki, 70, 240n15
air-conditioning, 33–34, 98, 177, 183, 248n89; in cinema theaters, 5, 16, 111, 187; used to filter out coal smoke, 184
Alexandra House, 33
Anslow, Barbara, 77
anti-Semitism, 4, 103, 104, 105
appliances, electrical, 110, 111, 146
Arthur, Prince (Duke of Connaught), 32
Aserappa, J. P., 147
Asian Industry magazine, 176
Asia Pictures Studio, 141
Asiatic Energy System, 70
Atlee, Clement, 89
Australia, 35, 40, 172, 178; CLP as investor in, 224; coal imports from, 179, 182
Aw Boon Haw, 130

Babcock & Wilcox, 41
Babington, Anthony, 42
Bailey, Patrick, 176, 177

Barber, Charles H., 115, 127, 136, 141–42, 148
Barker, J. W., 72–73, 172
Barlow, B. I., 147
"Beatnik," 140
Belilios, Emmanuel, 65
Bell, Alexander Graham, 31
Bell, George, 165
Below the Lion Rock (television series), 257n93
Benham, Harold Dudley, 103, 246n48
Benjamin, Kelly & Potts, 42
Benjamin, Sassoon, 42
Bennett, Charles J. M., 134, 135
Bernacchi, Brook, 130, 132, 140
Beveridge Report (1942), 89
Black, Governor Robert Brown, 99–100, 119, 182, 183
blackouts, 45, 77, 232n51
Black Point, gas-fired plant at, 215
Bombay, 8, 28, 173, 207; Crawford Market, 35; telegraphy in, 31
Boon, Goh Chor, 232n41
Borrowed Place, Borrowed Time (Hughes, 1976), 180
Braga, Noel, 50, 237n95

INDEX

Braga, José Pedro (J. P.), 54, 56, 122–23, 179, 243n102
Britain, 9, 10, 90; Central Electricity Board, 134, 251n44; CLP's electricity supply from, 11; colony's reliance on, 31; electricity as political issue and, 120; expiration of lease on New Territories and, 12; first electricity supply station, 235n37; government controls over electricity distribution in, 123; imperial decline of, 197; nuclear power in, 114, 116; postwar political changes in, 89; public concern about safety of electricity, 44; Winter of Discontent strikes (1977), 208; in World War II, 27
British and Chinese Corporation, 33
British Electrical and Allied Industries Research Association, 175
British Electricity Board, 172
British Electricity Council, 175
Buchanan, Alex, 171
building regulations, 87
Burgess, Claude, 93, 253n87

Cable & Wireless, 147, 148
Cairo Conference (1943), 78
Calcutta, 35, 47, 173
Calcutta Electric Supply Corp., 36
Calder Hall nuclear reactor, 112, 116, 249n102
Callaghan, James, 11, 181, 193, 208
Cameron, N., 74
Canton (Guangdong), city of, 41, 43, 88, 205; Chiang Kai-shek's power base in, 53; electrification of, 65; managing demand for electricity in, 45; Shameen (Shamian) district, 40, 44
Canton Electric and Fire Extinguishing Co., 42, 43
Canton Electric Light Works Co., 40
Canton Insurance Office Ltd., 64
capital investment, 14, 47, 55, 65, 123; Cowperthwaite's ideas on, 156; economies of scale and, 36, 121; Esso–CLP joint venture and, 163, 164, 166; in high-quality technology, 66; Kadoories' takeover of CLP and, 162; nationalization proposals and, 153
capitalism, 23, 24, 89, 159, 166; electrification and, 19; laissez-faire, 3, 15; managerial, 56
Caro, Robert, 44, 124
Cartwright, H. A., 44
Castle Peak electricity plant, 210, 221; British manufacturing and, 11; Thatcher's inauguration of (1982), 1–2, 5, 17, 193–94, 208, 211, 216, 221
Castle Peak Power Co., 168
CCP (Chinese Communist Party), 22, 53–54, 79, 222
Chan Chak, Admiral, 79
Chase National Bank, 92
Chater, Paul, 32, 37, 38, 51, 60, 65
Chater House, 33
Chen, Percy, 132, 139–40, 244n3
Cheong-leen, Hilton, 132, 140
Cheung Sha Wan Factory, 147, 148
Chiang Kai-shek, 53, 54, 78, 79, 88
Chiang Kai-shek, Madame, 67
China, People's Republic of (PRC), 11, 88; coal purchased from, 182; Communist victory in civil war, 69, 89, 91, 169; Cultural Revolution, 20, 23, 171, 199, 201; electricity sold to, 194, 197; expiration of British lease on New Territories and, 12; Great Leap Forward, 18; Hong Kong's post-1949 separation from, 19; imports from, 178, 179; Kadoorie China assets seized by, 206; Korean War and, 19, 92, 93, 166, 199; as market for CLP, 204–7; nationalization of industries in, 120; Tiananmen Square massacre (1989), 217; United Nations seat won back from Taiwan, 200; U.S. and UN embargoes against, 19, 92, 93, 166, 199, 204, 206
China, Republic of, 25, 81, 206. *See also* KMT [Kuomintang]
China and Manila Steamship Co., 41
China Resources, 181
Chinese Commercial Union, 46

INDEX

Chinese Exclusion Act (United States, 1882; extended 1904), 45, 170
Chinese Manufacturers Association, 142, 175
Chinese Reform Association, 90, 139, 244n3
Ching, Harry, 77
Choi Waihung, 244n3
Chu, C. C., 130
Churchill, Winston, 67–68, 78, 89
Churchill Falls hydroelectric facility (Canada), 209
Civic Association, 131, 132
civic reform associations, 7
civil war, Chinese, 53, 79, 104, 199
Clarke, Arthur, 104, 119, 127, 128, 149, 253n100
Clementi, Governor Cecil, 123, 125, 237n95
CLP (China Light & Power Company Ltd.): attempted nationalization of, 13; bonuses paid to staff for strike losses, 170, 255nn21–22; boycott against, 46, 170; British military takeover of (1945–1946), 90–91, 223; colonial state and, 15; Consulting Committee, 31; Cultural Revolution upheavals (1967–1968) and, 171–72; as debt-free company, 157, 254n112; in development of Hong Kong, 214–20; early years in Canton and Kowloon, 42–52; electricity pricing and, 57; electrification of New Territories and, 122–23, 125, 237n95; electrification of squatter areas and, 187–91, 257n88; ESCC and, 127–28, 137–41; Esso joint venture with, 20, 163–68, 195, 217–18; ethnic makeup of staff, 172; factories served by, 106, 247n57; founding of (1900), 42, 229n3; generating capacity of, 17, 61, 196; government administrators and, 13; government controls over, 4, 15–16, 24, 123, 220–22; idea of Hong Kong shaped by, 16; industrial and residential rate differentials and, 37; investments in Asia, 6; Kadoorie family control of, 120, 162; local management autonomy of, 12; mainland China as market for, 204–7; maintenance engineers, 72–73; merger proposal with HKE, 151, 160, 182, 220; nationalization prospect and, 210; opportunities and dangers (1950s–1960s), 100–108; as parallel state, 14; restart of operations after World War II, 82–88; restructurings of, 52, 60; service area of, 3, 17, 23, 109, 128, 248n76; special funds for cash reserves, 154; state-business relations and, 222–24; technology embraced by, 8–12, 30–31, 108–12, 174–77; wartime transformation of, 27–28
CLP board of directors, 18, 103, 160, 237n95; Kadoories as members, 54, 55; Kadoories' takeover of, 54, 56, 238n118
CLP critics, 137–41; Barber, 141–42; heads of resettlements, 147–48; Holmes report and, 142–46; Kadoorie and, 148–51; New Territories villagers' grievances, 143–44, 148
CLP shareholders, 42, 114, 127; consumers' funds used to finance expansion, 155; dividends paid to, 154–55; ESCC report and, 152, 253n87; Kadoories as controlling shareholders, 25, 158, 223, 254n114; merger proposal with HKE and, 157–58; nationalization prospect and, 160; original shareholders, 54; profits distributed to, 58; property rights of, 13; resumption of Chinese sovereignty and, 208, 209; SoCA and, 164; value of CLP compared with HKE, 49
coal, 1, 9, 50, 230n10; British imperial policy and, 181; Castle Peak plant and, 2; Hong Kong's lack of natural resources and, 9; imported from China, 179, 208; Japanese occupation and, 75–76, 241n55; nuclear power as threat to, 112, 113; postwar increase in cost of, 243n91; price of, 45, 179–80; switch to oil at generating stations, 164, 174

INDEX

Coates, Austin, 38, 74, 231n38, 232n42
Cold War, 23, 91, 92–93, 139. *See also* geopolitics
colonial state, 15, 221; crime-fighting and, 22; private business and, 69; responsibilities of, 119; "system trust" and, 238n111
Committee for a Free Asia, 24
communism, 23, 139
Communist Party, Chinese. *See* CCP (Chinese Communist Party)
Companies Act (Hong Kong, 1865), 65
compradors, 41, 46
computers, 109, 176, 187
Convention of Peking (1860), 229n3
Cowperthwaite, John, 4, 5, 12–13, 118, 119, 131; attempt to force merger of electric companies, 13, 152, 155; corporate-state cooperation and, 14, 137; on dangers of monopoly, 155; dividend restrictions proposed by, 220, 253n87; economic interventionism of, 126, 153–54; nationalization supported by, 159; on shareholder use of consumers' funds, 155, 156–57, 158, 159; state controls over supply companies and, 160; upheavals (1967–1968) and, 171–72
Cox, A. R., 74
Cox, R. G., 146
Crew, Leonard, 246n51
Cutler, David, 234n23

Daily Information Bulletin, of HK government, 133
Dairy Farm, 177–78, 235n48
Danby, Leigh & Orange, 41
Danby, William, 41
Dairen (Dalian), city of, 36
Davis, Governor John, 21
Daya Bay nuclear complex (China), 116, 182, 193, 194, 210, 215
Delano, Warren, 46
demand, for electricity, 44, 52, 64, 66; capacity built in advance of, 2, 58, 61, 100, 219; CLP revenues and, 16; cost of new generating equipment and, 121; economic growth and, 19; electricity pricing and, 47, 56–57; from factories, 24; under Japanese occupation, 76; management of, 45, 48; surges in, 57; tramways and, 37; water pumped from reservoirs and, 39; wealth and, 8
democratization, 125, 249–50n18
Deng Xiaoping, 20, 182, 198, 216, 258n14; meeting with Thatcher (1982), 1, 192, 201, 211; rise to power, 200; on status of Hong Kong, 202
Denkisho (Japanese occupation electricity authority), 75
Department of Public Works, 123
depreciation, 58, 113, 114, 154, 254n112
Des Voeux, Governor William, 31, 32, 35
Dietz, Frederick, 51
disease, 93, 94, 99, 100
dividend payments, 66, 103, 104; amounts in 1950s, 155; based on paid-in capital, 58, 102; Cowperthwaite's proposals and, 153, 154, 156, 158; easing of restrictions on, 163; government restrictions on, 13, 24, 86, 102, 117, 122, 161, 220; nationalization and, 151; profits paid in form of, 105, 122; reliability of, 36
Dublin, introduction of electric light in, 8
Duff, James, 42

East Asian miracle, 19
economic development, 4, 20, 159, 215; of China, 199; role of electricity in, 3, 213, 226
economies of scale, 36, 47
Edison, Thomas, 34, 48, 124, 172–73
Edwards, W. S., 140–41
Eisenhower, Dwight, 22, 23, 166
Electrical Ordinance (1911), 122
electricity: in colonial Asia, 34–41; competition for in-house generation of, 47; early opponents of, 43; economic development and, 3, 213, 226; economic growth and, 63, 69; electricity supply as fundamental

INDEX

right, 125; everyday life shaped by, 20, 91; idea of Hong Kong and, 15–24; idea of the city and, 183–87; industrial growth and, 2; introduced to Hong Kong (1890), 8; Japanese occupation and, 70–78; limits of Hong Kong laissez-faire and, 12–15; modernity/modernization and, 3, 5, 17, 65, 101, 184–85, 187, 215, 224; naturally produced, 7; as natural monopoly, 120, 121, 160; political struggles over payments for, 2–3; public debate in newspapers about, 139; state control and, 90, 100, 102, 117, 118; as technological system, 6–8; uneven diffusion of, 51–52; wealth production and, 7–8

electricity, as political project, 90, 117–21; formation of ESCC, 127–34; state regulation of electricity, 122–26

Electricity Act (India, 1903), 124

electricity distribution, 6, 20, 48; after World War II, 83; Asia-Pacific distributed supply network, 28; in Canton, 44; natural monopolies and, 121; terminal points of, 14

electricity meters, 22, 31, 125, 175, 188–89

electricity pricing, 3, 16, 39, 104, 139; differential rates for consumer versus bulk industrial usage, 37, 57, 149; fuel surcharge (1941), 71, 240n12; postwar rate increase, 85–86; resentment among rural customers over, 145; state control and, 118, 124

Electric Lighting Acts (Britain, 1882, 1883), 123

electrification, 2, 15, 137; as instrument of power, 29; of squatter areas, 187–91; in the United States, 15, 44, 90, 124, 143, 173

elevators, 5, 22, 29, 183, 224

Elizabeth II, Queen, 112, 185

Elliot, Charles, 26

Endacott, G. B., 230n11

England, Vaudine, 82

escalators, 5, 22, 173–74, 183

ESCC (Electricity Supply Companies Commission), 11, 24, 47, 146, 172; aftermath of hearings, 161; Barber testimony, 115; complaints about supply companies and, 138; formation of, 119, 127–34; hearings (1959), 84–86, 104, 118, 122, 139, 150, 162, 198, 210, 220, 221–22, 224; membership of, 134–37; merger and nationalization recommended by, 117; nationalization recommended by, 157; report following hearings, 151, 160, 253n87

Esso oil company, 5, 10, 58, 116, 120, 161–62, 176, 210; investment in Castle Peak, 216; joint venture with CLP, 20, 163–68, 217–18; SoCA and, 163–68; as world's largest oil company, 161

Europe, 15, 31, 120

Exxon Energy Ltd., 168

Faber, S. E., 61

Faraday, Michael, 34–41

Far East American Enterprises Ltd., 136

Far Eastern Economic Review, 114–15

Far Eastern Review, 237n93, 256n55

Financial Times, 165, 168, 194

fire danger, 22, 45, 95, *101*, 189; electricity code and, 146; fire pumping and firefighting, 42, 45, 190–91; kerosene and, 184; in new towns, 23; Shek Kip Mei fire (1953), 24, 94, 95, 119, 219, 245n25; Tai Hang Tung shantytown fire (1954), 94; Windscale nuclear facility fire (Britain, 1957), 114

fishing industry, electricity and, 9

Fish Marketing Organization, 80

Forsyth, Hector, 147–48

Fortescue, Tim, 81

fossil fuels, 2, 115, 162, 174, 179, 217

France, 9, 116

Fraser-Smith, Robert, 38–39, 40, 174

Friedman, Milton, 12–13, 159

fuel surcharges, 71, 130, 250n32; Holmes's recommendation to abolish, 145; unpopularity of, 24, 125

Fung, Victor, 203

Fung Wa-chuen, 41, 42, 46

INDEX

Gang of Four, 199–200
Gavriloff, George, 72, 86
"General Policy (A Brief Survey)"
  (L. Kadoorie), 56, 57–59, 148, 239n122
geopolitics, 11–12, 20, 22, 91, 92, 214;
  Castle Peak electricity plant and, 2,
  215; Chinese refugees in Hong Kong
  and, 93; Esso investment and, 165,
  166. *See also* Cold War
Germany, 9, 27, 64
Gerschenkron, Alexander, 71
Gimson, Franklin, 78, 242n60
global celebratory events, electricity
  used in, 185
Goldman, Stephen, 168
Goodstadt, Leo, 128, 131
Gordon, Sydney, 156
Graham-Cumming, G., 146
Grantham, Governor Alexander, 18, 69,
  81, 100, 128, 232n47; on the Chinese
  Reform Association, 244n3; on
  expiration of New Territories lease,
  92; policy of subsidizing industry
  and, 149; as private enterprise
  supporter, 119; "problem of people"
  phrase, 93; report on 1956 riots and,
  95, 96; on trade as lifeblood of Hong
  Kong, 108
Grayburn, Vandeleur, 33
Great Depression, 33, 70, 125, 197
Green Island Cement Co., 40, 60, 180
Guangdong Province (China), 88, 162,
  205, 215; Castle Peak plant and, 194;
  economic growth in, 216; electricity
  production in, 42–43; Japanese
  invasion of, 27, 56, 67, 96, 179, 206;
  open and closed border with, 96;
  supplies sent to Hong Kong from, 93,
  182. *See also* Daya Bay nuclear
  complex

Haddon-Cave, Sir Philip, 12
Halpern, Eric, 114–15
Hankow (Hankou), city of, 36, 41
Harcourt, Admiral Cecil, 79, 82, 84
Hayes, General Eric, 83
Heath, Edward, 200, 258n14
Heller, Thomas, 42, 43
Hemingway, Ernest, 67
HKE (Hong Kong Electric), 2, 8, 11, 25,
  85, 138, 199; Albany pumping station
  and, 37; Ap Lei Chau generating
  facility, 185; attempted merger with
  CLP, 13; attempted takeover of CLP,
  55; British military takeover of
  (1945–1946), 90–91; coal supplies of,
  179; difficulties supplying light and
  power, 105; Electrical Ordinance
  (1911) and, 122, 124; electricity pricing
  and, 57; electrification of New
  Territories and, 122–23, 125; ESCC
  hearings and, 128; first street lights in
  Hong Kong and, 30; formal contract
  with government, 139; history of
  unbroken management, 230n17;
  Hong Kong Island served by, 3–4;
  local management autonomy of, 12;
  mainland China as market for,
  204–5; merger proposal with CLP, 151,
  160, 182; modernity associated with
  electricity, 184–85; North Point
  generating station, 73; officers of, 41;
  as publicly traded company, 49; share
  prices of, 133; size of, 17; technology
  and, 110, 174, 175
HKHP (Hong Kong Heritage Project),
  110, 146, 176, 196, 197, 230n13
Hok Un (Hok Yuen) facility, 48, 59–66,
  105, 154, 162, 233n10; depreciation
  charges and, 113; Japanese invasion/
  occupation and, 70, 72, 75, 76, 77;
  Kadoorie's speech at opening
  ceremony (1940), 25, 26, 62, 206, 217;
  location on map, *101, 204*; planning
  and construction of, 26; relations
  with PRC and, 206; repair and
  expansion of, 87, 106–7; restarted
  after the war, 82; technologies and,
  108, 109, 112, 176; World War II
  and, 27
Holmes, D. R., 136–37, 142–46, 148
Hong Kong: as "barren rock," 26, 62; as
  British Crown Colony, 1, 3–4, 11,
  236n56; British empire serviced by,

INDEX

169; closed border with China, 96, 97, 207; economic acceleration (post-1945), 2, 5, 169, 214, 259n1; Esso investment in, 5; existence of two electricity companies in, 4; expanded state role in economy, 69–70; film studios, 24, 141; "industrial revolution" of, 169; lawless reputation in 19th century, 21–22, 231n35; limits of laissez-faire economy in, 12–15, 118, 126, 222–23; maps, *101, 204*; modernization enabled by CLP electricity, 214–20; night passes for Chinese residents in colonial era, 21; population figures, 16, 27, 70, 91, 93, 219, 240n9; post-1945 society made by electricity, 16–24; resumption of Chinese sovereignty (1997), 1–2, 13, 192, 197, 213, 217; as Special Administrative Region in PRC, 202, 206–7, 218; technology in, 8–12; as textile producer, 5, 16, 169; as "West Berlin of the East," 23, 91, 93, 200. *See also* Kowloon; New Territories
Hong Kong, Japanese invasion and occupation of, 67–68, 96, 179–80; British return after Japanese defeat, 78–82, 180; electricity during, 70–78
Hongkong & Yaumati Ferry Co., 126
Hong Kong and China Gas Co., 30, 37
Hongkong and Shanghai Bank. *See* HSBC
Hong Kong and Whampoa Dock Co., 26, 47, 59, 86, 237n86; Kadoories' investments in, 60; World War I and, 52
"Hong Kong—Boom Town" (*Collier's*, 1948), 97–98
Hong Kong Club, 33, 81
Hong Kong Cotton Spinners Association, 130
Hong Kong Engineering & Construction Co., 61, 110
Hong Kong General Chamber of Commerce, 9
Hong Kong Island, 32, 42, 68, 73, 183, 192
Hong Kong Jewish Refugee Society, 103

Hong Kong Polytechnic University, 30
Hong Kong Rope Manufacturing, 107
Hong Kong Telephone, 126, 135
Hongkong Tiger Standard (*HKTS*), 128, 129, 130, 135, 152
Hongkong Weekly Press, 43–44, 49
housing, 14, 16, 91, 98, 119, 221; demolished in Kowloon Walled City, 81; public housing estates, 24, 147, 219, 222, 245n25; refugees and demand for, 64, 89, 93–94; Sheung Li Uk project, 245n25; wartime destruction of, 70
Hou Tin-sang, 191
HSBC (Hongkong and Shanghai Bank), 33–34, 55, 80, 111, 157
Hughes, Richard, 180
Hughes, Thomas, 15, 24
Hume, G. A., 189
Hydro-Quebec, 120

ice, electricity in production of, 9, 177–78, 183, 256n50
India, 29, 40, 90, 124, 180
Indochina, French, 179, 180, 230n10
Indonesia, 126, 180, 182
Insull, Samuel, 48, 58, 124, 143, 157, 222
International Combustion Ltd., 61
Internet, 7

James, Weldon, 98, 245n35
Japan/Japanese empire, 9, 40, 179, 230n10; invasion of China (1937), 26–27, 56, 179, 206; occupation of Hong Kong, 70–78
Jardine, Matheson & Co., 33, 81, 106, 230n11
Jews: ambiguous social position in Asia, 54; Baghdadi (Mizrahi) Jews in East Asia, 103, 238n112; refugees from Nazism in Hong Kong, 86, 243n96
Jiang Qing, 199
Joffe, Eugene, 73, 74, 144, 170
Johnson, Lyndon Baines, 44
Johnson & Phillips, 41
Jones, Raymond, 77
Jordan, J. H., 33–34

INDEX

Kadoorie, Eleazer ("Elly") Silas, 29, 42, 243n102; in CLP corporate archive, 54; early life of, 52–53; early years of CLP and, 44; electrification of New Territories and, 122–23, 237n95; Hong Kong Club membership of, 82; rushed British naturalization process of, 53, 59, 64, 238nn111–12

Kadoorie, Ellis, 52, 82

Kadoorie, Horace, 42, 54, 103; as chairman of Shanghai Gas, 238n118; diary shared with Lawrence, 56, 59, 157

Kadoorie, Lawrence, 7, 17, 20, 29, 214; on boycott against U.S.'s Chinese Exclusion Act, 45, 46; British civil service connections of, 11; Castle Peak plant and, 1, 208, 210; as chairman of CLP, 4; on China as market for CLP, 205–6, 216; corporate-state cooperation and, 14, 137; critics of CLP and, 148–51; denial of monopoly status for CLP, 127–28, 148; on distribution in Canton, 44; early life of, 206–7; economics of electricity supply and, 48, 58; electricity viewed as politically neutral technology, 65; electrification of squatter areas and, 191; at ESCC hearings, 148, 149, 153, 187; Esso joint venture and, 10, 164–66, 210; on establishment of CLP, 42; expiration of New Territories lease and, 69; fears of economic collapse, 209–10; first associations with CLP, 52; formation of ESCC and, 131–34; government interference and, 117; on growing demand for electric current, 45; as heroic entrepreneur and visionary, 170; Hok Un expansion and, 54, 61, 62–63, 110, 217; Hong Kong Club membership of, 82; Japanese invasion/occupation and, 73, 74; manufacturing in Hong Kong and, 17–18; meetings with British politicians, 223; merger proposal and, 155; Mizrahi Jewish heritage of, 4; Nanyang Cotton Mill and, 107; nationalization prospect and, 152–53, 157, 170, 187, 203; nighttime illumination and, 23, 232n51; "1997 and All That," 195–204, 208, 211–12; nuclear power and, 112–16, 182, 193, 205, 215, 216–17; peerage status of, 11; political and business network of, 11; postwar restart for the electric city, 82–88; public image of CLP as concern of, 148; relations with PRC and, 92, 166–67, 182, 201–2, 205–7, 257n65; resumption of Chinese sovereignty and, 2, 195, 196, 212; technology embraced by, 8, 9, 174, 255n36

Kadoorie, Michael, 84, 106, 164

Kadoorie family, 4, 25, 42, 48; CLP and, 52–59; companies invested in, 60; imprisoned under Japanese occupation, 68

Kai Tak airport, 32, 92, 99, 129, 219–20

Kao Sun Zee, 172

Kau Wa New Village, 144

Keswick, John, 81, 98, 106

King, Frank, 87

KMT [Kuomintang] (Nationalists), 23, 78, 79, 85, 88, 130; Chinese civil war and, 22, 221–22; triad gangs and, 95. *See also* Chiang Kai-shek; China, Republic of

Koo, Wellington, 78

Korea, 70

Korea, North, 93, 167

Korea, South, 126

Korean War, 19, 92–93, 166, 199

Kowloon, 26, 49, 105, 229n3; as "barren" area, 26; CLP as electricity provider to, 3; decorative lighting in Mid-Autumn Festival, 183; factory owners subsidized by farmers and workers, 152; gas lighting in, 50–51, 237n100; pro-Japanese Fifth Column elements in, 72–73; Qing cession to Britain (1860), 192; St. Andrew's Church, 51; street lighting in, 14; Tsim Sha Tsui area, 183, 186

INDEX

Kowloon-Canton Railway, 26, 31, 33, 59, 60
Kowloon Walled City, 22, 81, 101, 202
Kwangtung Electric Supply Co., 41
Kwanti Substation, 147, 148
Kwok, S. T., 76
Kwun Tong industrial complex, 129

Lai Chi Kok, 10, 161, 248n89
Lancaster, Richard, 239n122
Landale, David, 114
Lane Crawford company, 39
Laufer, Edgar, 86, 138
Leet, Charles, 163, 167
Legislative Council (Legco), 55, 99, 131, 134, 141; Braga and, 54; Grantham's address to, 108; Lawrence Kadoorie as member of, 96
Lenin, Vladimir, 18
Lennox-Boyd, Alan, 95
Lent, John, 130
Lévesque, René, 120
Li Cheng Uk, 147
Lichtrequisit einer elektrischen Bühne [light prop for an electric stage] (Moholy-Nagy, 1930), 186
Lightbody, I. M., 188
light/lighting, 231–32n38; everyday life shaped by, 29; gas lighting, 30, 50–51, 234n15, 237n100; state order associated with, 21; street lighting, 43
Li Han-chang, 43
Lin Biao, 199
Lion Rock, 189
Li Shu-fan, 75–76
Lo Cheung-shiu, 250–51n35
Lo Man-kam (M. K. Lo), 103, 131, 170, 250–51n35
London, city of, 8, 24, 29, 173

Macao, 67, 98, 200, 201
Macartney, George, 192
MacDougall, David, 79, 80, 83, 85, 102
MacLehose, Governor Murray, 92, 220
Macmillan, Harold, 23, 141
Malaysia, 126
Malloy, Flight Lieutenant, 84

Ma Man-fai, 132–33, 140
Manchukuo, 70, 241n55
Manchuria, 26
Mandarin Hotel, 33
Manila Electric Railroad and Light Company (Meralco), 37, 235n47
Mao Zedong, 92, 171, 199–200, 203, 205; meeting with Heath (1974), 200–201, 258n14; portrait in Tiananmen Square, 1
Marx, Karl, 15
Mei Foo Investments, 248n89
Melbourne, 35, 47, 173
Metropolitan-Vickers Export Co. (Manchester), 61, 62, 87
mining, 105, 180, 182, 208, 256n60
Mizrahi Jews, 4
Mobil oil company, 10, 161, 165, 217
modernity/modernization, 3, 16, 17, 23, 65, 112
Mody, Hormusjee, 65
Moholy-Nagy, László, 186
monopoly, 58, 120, 121, 152, 160
Moore, James H., 248n89
Morgan, E. Wilmot, 138
Morse, Arthur, 87, 114
Mould, John, 134, 135, 136
Municipal Council of the International Settlement (Shanghai), 36
Munton, D. W., 84

Nanyang Cotton Mill, 107
Nanyang Textiles, 18
nationalization, of electricity supply, 11, 13, 57, 102; accommodation as alternative to, 151–59; in Britain, 119; buyout of shareholders and, 160; Clarke as supporter of, 127, 128; CLP's resistance to, 24; Cold War and, 139; ESCC and, 117, 157; Esso investment and end of discussion about, 168; merger of supply companies and, 151, 160, 182, 220; in Quebec, 120
New Territories, 2, 39, 105, 222; British lease of (1898), 2, 22, 26, 54–55, 78, 101, 193, 229n3; CLP as electricity provider to, 3, 23, 54, 237n95; electricity pricing

INDEX

New Territories (*continued*)
  in, 139; electrification of, 122–23, 125, 136, 142–43, 237n95; expiration of lease on, 11–12, 91, 92, 149, 194, 197; factory owners subsidized by farmers and workers, 152; Kowloon-Canton Railway and, 31, 33; street lighting in, 14
New York City, 8, 29, 134, 173, 183
Ngo, Tak-wing, 107–8
"1997 and All That" (L. Kadoorie), 195–204, 208, 211–12
Nixon, Richard, 200
Noble, Joseph, 31
Northcote, Governor Geoffrey, 25, 26, 27, 63
nuclear power, 9, 112–16, 129, 134, 182, 211. *See also* Daya Bay nuclear complex
Nye, David, 7

Ocean Terminal, 186, 220
oil, 9, 80, 164, 174; oil reserves, 181; oil shock (1979), 181; OPEC embargo (1973), 209; success of oil-fired generation, 181
"One Country, Two Systems" policy, 200, 201
Onslow, Alan, 172, 176
OPEC (Organization of Petroleum Exporting Countries), 120, 209
Opium Wars, 26, 63, 106, 192
Osterhammel, Jürgen, 28

Palmerston, Lord, 26
Paris, city of, 24, 29
Parsis, 33, 65, 134
Paterson, Major J. J., 73
Peak Tram, 31, 32, 47, 235n48, 246n47
Pearl River Delta, 205
Pedder, R. A., 189
Peng Chau Rural Committee, 231n21
Peninsula Hotel, 67, 68, *101*, 111, 186, 221
PEPCO (Peninsula Electric Power Co.), 167–68, 195, 197
Petersson, Niels, 28
petroleum companies, multinational, 10
Philippines, 37, 126

Pickering, J. W., 165
Po Leung Kuk, 41
Po Yick Association, 140
Praya Reclamation Project (1890s), 32–33, 219
PRC. *See* China, People's Republic of (PRC)
Preece, Cardew & Rider, 31, 61, 64
price controls, 79, 85, 90, 98
Problem of People, A (government report, 1957), 95, 99
Project '77, 208–12
property rights, 13, 65, 99, 126, 154, 216
public health, electricity and, 22, 34
public–private joint projects, 15

Qing dynasty, 192, 199
Queen Elizabeth Hospital, 99, 219–20

radio, 32, 109, 136, 174, 185, 187
Raymond, Abraham Jacob, 246n47
Raymond, Albert, 82, 102, 103, 246n47
Reform Club of Hong Kong, 90, 128, 130, 131, 132, 141, 244n3
refrigeration, 29, 124
refugee population, 18, 23, 27, 69, 70, 89; housing for, 64, 93–94; numbers of, 96, 101; as problem for colonial authorities, 93; state efforts to impose urban order and, 100
Reid, Archibald, 42
Report on Electrification of Squatter Areas (1976), 250n20
Report on the Riots in Kowloon, 95
reservoirs, 14, 99
Reuter, Julius, 31
Reuters news agency, 31, 234n23
Reyrolle [A.] & Co. Ltd., 64, 82, 172
Ribeiro, G.H.V., 172
riots (1967), 171
riots (Kowloon, 1956), 23, 95–96, 99, 119, 232n47
Roberts, Priscilla, 166
Rockefeller, John D., 161
Rolfe, C. S., 172
Rong Yiren, 107, 247n65
Roosevelt, Franklin Delano, 46, 78

INDEX

Royal Air Force, 82, 107, 180, 190
Royal Navy, 72, 79, 84
Rural Electrification Administration, 124, 143
Russell & Company, 45–46
Ruttonjee, Dhun Jehangir, 134–35
Ruttonjee, Jehangir, 134–35
Ryrie, Phineas, 37, 235n48

safety inspections, 147, 190, 247n57
St. George's Building, 23, 28, 33, 52, 171
Sakai Takashi, Lieutenant, 68
Sandakan Light & Power Co., 41
Sassoon, Victor, 103
Sassoon family, 4, 53, 65, 103, 115
Schivelbusch, Wolfgang, 21
Schwartz, Margaret, 167
Sham Shui Po, 68, 245n25
Shanghai, city of, 8, 18, 53, 149–50, 187, 207; factories moved to Hong Kong, 17, 18, 107, 214; International Settlement, 132; Marble Hall mansion of Kadoories, 55, 82, 206; uneven development of modernity in, 51–52
Shanghai Electric Co., 36
Shanghai Gas Co., 42
Sha Tin, town of, 23
Shaw Studio, 141
Shek Kip Mei fire (1953), 24, 93, 95, 119, 219, 245n25
Shek Wu Hui Chamber of Commerce, 142, 145
Shenzhen (Shum Chun), town of, 27, 182–83
Shewan, Robert, 8, 31, 40, 65, 159; on coal consumption for generators, 45; on cost of coal, 179; death of, 55, 239n120; founding of CLP and, 42, 48–49; interest in electricity, 43; Kowloon generating plant and, 49
Shewan Tomes & Co., 28, 31, 60, 162, 233n10; as agent for Metropolitan-Vickers, 87; businesses managed by, 40–41, 107; CLP run as sideline by, 65; electrification of New Territories and, 123; investment in wide range of businesses, 43, 55

shopping malls, 5
Singapore, 8, 31, 127, 173, 232n41, 234n15
Sing Tao newspaper chain, 130
Sinn, Elizabeth, 96
Snow, Philip, 68, 81
SoCA (Scheme of Control Agreement), 5, 120, 145, 150, 183, 208, 223–24; benefits for CLP of working with government, 198; Esso-CLP joint venture and, 163–68, 218
Soong, T. V., 88
Sorby, Vincent, 73, 74
South China Morning Post (*SCMP*), 130, 133–34, 135, 152, 173, 250n31
squatter settlements, 18, 95, 163; CLP employees' ventures into, 101; electricity provided to, 14, 22; illegal tapping of electricity in, 250n20; Shek Kip Mei, 24; working party on electrification of, 187–91
Standard Oil trust, 161, 217
Stanley Internment Camp, 68, 77, 78, 81, 135, 241n49, 242n60
Star Ferry Co., 31, 102, 126, 127
Steele, Colin, 189
Steinschneider, Paul, 86, 243n96
Stephens (Reyrolle & Co. Ltd. director), 82, 88, 206, 243n107
Stoker, William, 120, 131
Straits Settlements, 40, 41
streetcars, electric, 31
Suez Canal, 9, 29
Summary of Submissions (Holmes), 137, 142
Swan, Joseph, 34
Szczepanik, Edward F., 139, 175, 247n58

Tagg, G. T., 131
Tai Hang Tung, 147
Tai Hang Tung shantytown fire (1954), 94
Taikoo Sugar, 50
Tai Lam Chung reservoir, 99
Tai Loo Ling village, 144
Tai Ping Carpets International, 107
Tai Po District, 14, 142–43
Taiwan, 23, 126, 200, 201
Tai Yuen village, 250n20

tariffs, 16, 24, 58, 103, 125, 139, 223; British military administration at war's end and, 123; price of coal and, 85; reduced rates, 160, 163; shareholders' dividends and, 58

Taus, E., 111

Taylor, Elizabeth, 104

technology, 6, 30–34, 91, 162; as civilizing force, 39; CLP's uses of, 108–12; communication, 108, 109, 174–75; corruption and instability of, 62; general purpose versus niche, 7; globalization of, 28–29; modernity and, 9, 185, 218; rapid introduction into Hong Kong, 172–78; state control and, 95–96; transformative impact of electricity and, 213

Teesdale, Edmund, 253n87

telegraphy, 9, 29, 31, 51, 173, 230n11

telephone networks, 174–75

television, 110, 183

Tennessee Valley Authority, 124, 143

Thatcher, Margaret, 11, 20, 192–93, 203; Castle Peak plant inaugurated by, 1–2, 5, 17, 193–94, 208, 211, 216; denationalization program of, 194–95; visit to Hong Kong (1982), 1, 192, 201, 211, 213

Tientsin (Tianjin), city of, 36, 41

Todd, Michael, 104

Tomes, Charles, 45

transportation systems, 5, 102, 215

Trans-Siberian Railway, 59

Treaty of Nanking [Nanjing] (1842), 4

treaty-port system, 55, 78, 206

Trench, David, 136, 142, 146

triad gangs, 72, 95

Tsang, Steve, 80, 107

Tsingtao (Qingdao), city of. 36

Tsing Yi electricity plant, 10, 164, 168, 171, 195; construction of, 176; oil prices and, 181. *See also* PEPCO

Tu, Elsie Elliot, 140

Turner, Michael, 253n100

Twentieth Century Impressions (Wright and Cartwright, 1908). *See* Wright and Cartwright

United Nations, embargo on China (1951), 19, 92, 199, 206

United Nations Association of Hong Kong, 132

United States, 10, 11, 24, 31, 43; Chinese Exclusion Act, 45, 170; coal purchased from, 182; electricity as political issue and, 120, 121; electrification in, 15, 44, 90, 143, 173; embargo imposed on China (1950), 19, 92, 93, 166, 204, 206; Hong Kong's growing dependence on, 20; as market for Hong Kong manufactures, 97; public concern about safety of electricity, 44; stake in future of Hong Kong, 166; state regulation of electricity supply in, 124; in World War II, 27, 67, 77, 78

University of Hong Kong, 30, 75

Urban Council, 132

urbanization, 64

Ure, Colin J., 173–74

Vegetable Marketing Organization, 80

Victoria, city of (Central Hong Kong), 32, 49

Victoria, Queen, 35, 234n15

Victorian Electric Company, 35

voltage, 111, 138, 174, 175; fluctuations in, 146, 147, 148; high-voltage equipment, 8, 83, 185, 219; reduction of, 144

Wadar Motion Picture & Development, 141

Wang, Y. C. (Wang Yuncheng), 18, 106

water supplies, 69, 92, 178–79; from PRC, 182–83; shortages, 18, 68, 93

Watson (A. S.) company, 39

Weiss, Karel, 114

West Berlin, Hong Kong compared to, 23, 91, 93, 200

Westinghouse company, 43, 49

Westphal, Fred, 210

Whitelegge, David, 135, 136, 147

Wickham, W. H., 41

Winther, Tanja, 232n44

Wong, Wilfred, 111
Wong Ping-shueng (Huang Ping-ch'ang), 43
Wong Siu-lun, 131, 238n111, 247–48n72
Wong Tai Sin, 147
Wood, Cyril, 144, 171, 172
Working Party on Electrification of Squatter Areas, 187–91
World Bank, 158
World War I, 52, 238n112
World War II, 27, 69, 89, 221, 222. *See also* Hong Kong, Japanese invasion and occupation of
Wright, Arnold, 44
Wright, Frank Lloyd, 186
Wright and Cartwright, 40, 50, 52, 59–60

Wu, Gordon, 203
Wuhan, 43

Yan Wo Opium firm, 41
Yip Wai Ying, 191
Young, Governor Mark, 68, 81, 90, 132, 220
Yuan, L. Z., 130
Yuen Long, town of, 23
Yung Chi-kin, Larry (Rong Zhijian), 247n65
Yung Hwa Motion Pictures Studio, 141
Yunnan railway, 33, 230n10

Zhang, Chi, 42, 43
Zhao Ziyang, 193, 259n2
Zhou Enlai, 199, 200, 258n14